Introduction to Linear Programming: METHODS AND CASES

Introduction to Linear Programming: METHODS AND CASES

Thomas H. Naylor
University of Wisconsin

Eugene T. Byrne
Southern Methodist University

John M. Vernon
Duke University

Wadsworth Publishing Company, Inc.
Belmont, California

ISBN-0-534-00024-X

L.C. Cat. Card No.: 79-151721

Printed in the United States of America

1 2 3 4 5 6 7 8 9 10—75 74 73 72 71

Preface

This book introduces linear programming to individuals interested in business administration, economics, industrial management, and industrial engineering. It combines the theoretical and practical aspects of linear programming and provides a transition from simple problems to the complex problems of the real world. It also contains introductory chapters on input-output analysis and game theory.

For the reader whose background includes only a modicum of linear algebra and whose primary interest is the practical application of linear programming, this book offers sufficient theory to make practical application possible. For the reader who has had an abundance of mathematics, it provides a basic introduction to the simplex method and its application to industrial problems. In addition, those readers who wish to concentrate on the mathematical approach to linear programming may use it as a casebook.

The objectives of this book are twofold. The first is to present the simplex method for solving linear programming problems in a way that permits the non-mathematically inclined reader to understand the basic theory. This should enable the reader to solve, with a minimum of effort, any practical management problem that can be formulated as a linear programming problem. The second objective is to present a representative sample of problem situations, taken from business and industry, which can be solved by linear programming techniques.

The introductory chapter of the book attempts to clarify the relationship between linear programming and managerial decision-making. The introductory chapter is followed by a chapter that surveys linear programming techniques. Chapter 3 discusses the graphical approach to linear programming and serves as an introduction to the simplex method.

Chapter 4 should provide the reader with a good mathematical foundation for understanding the simplex method presented in Chapter 5. The development of the simplex method includes both a profit maximization problem and a cost minimization problem. Chapter 6 expands the discussion of duality which is introduced in Chapter 5. The importance of sensitivity analysis, once a solution has been determined, is also explained in Chapter 6.

Two special cases of linear programming, the transportation and assignment problems, are presented in Chapter 7. Chapter 8 considers how to solve linear programming problems when all variables are restricted to integer values. Input-output analysis is described in Chapter 9 and is illustrated by an example forecasting model. The final chapter is an introduction to game theory and its usefulness in analyzing conflict-of-interest situations.

It would be impossible to acknowledge all of the assistance that we received in writing this book; however, we do wish to mention certain individuals and organizations. Howard G. Schaller and C. Jackson Grayson of the School of Business Administration and W. David Maxwell of the Department of Economics, all of Tulane University, provided both inspiration and guidance. Daniel B. Killeen and Wilson J. Nettleton, Jr., of the Tulane Computer Center, provided assistance in solving the cases on the computer. The following business organizations provided assistance in preparing the case material: International Business Machines Corporation, TACA International Airlines, New Orleans Green Coffee Association, New Orleans Dock Board, and International Paper Company. We are indebted to the many professors who offered valuable criticism during the preparation of the manuscript. Finally, we would like to thank Robert Townsend, a student at Duke University, who read the entire manuscript and offered many helpful comments on it.

Thomas H. Naylor
Eugene T. Byrne
John M. Vernon

Contents

2
A Survey of Linear Programming Methods, 16

3
The Graphical Solution, 32

4
Simultaneous Linear Equations, 45

Linear Programming Methods

1 Linear Programming

The primary objective of this book is to acquaint students of management and economics with the mathematical technique known as linear programming. Two chapters are included with the secondary objective of introducing the related mathematical techniques of input-output analysis and game theory. To clarify these objectives, we wish to point out that this book is only slightly concerned with the mathematics itself. Instead, we present the techniques in a manner that allows the student whose background is essentially nonmathematical to comprehend the mechanics of solution and to recognize situations in which the techniques are applicable.

HISTORICAL BACKGROUND

Although linear programming, input-output analysis, and game theory are relatively new techniques, all of which have a common mathematical basis, they were developed independently. The common mathematical basis was not recognized until later. It is sufficient to state at this point that each technique is concerned with solutions to simultaneous linear equations.

Game theory and input-output analysis were developed first. Although John von Neumann published the central theorem of game theory in 1928, it was not until von Neumann and Morgenstern jointly published the *Theory of Games and Economic Behavior* in 1944 that game theory really had its beginning. As we shall see in Chapter 10, game theory is a mathematical approach to conflict-of-interest situations.

Input-output analysis is primarily the result of the research of W. W. Leontief, who was interested in describing empirically the production interdependence of goods and services in an entire economy. His first description of the method was published in 1936.[1]

Linear programming, the main topic of this book, was not developed until 1947, although mathematical techniques closely related to linear programming can be traced back much further. George B. Dantzig in that year published the complete description of the technique as we know it today.[2] Dantzig developed linear programming as a tool for solving planning problems of the United States Air Force.

Although linear "programming" and computer "programming" are completely unrelated as such, computer development has been extremely important for the success of linear programming. The reason for this is that linear programming is a technique for solving a problem by iteration. That is, the linear programming technique requires that a system of linear equations be solved over and over, with one variable being changed each time, until the final solution is obtained. An industrial example might involve 300 variables and 200 equations. To reach the final solution might require solving the system simultaneously, say, 100 times. Clearly, without a high-speed computer to perform the calculations, a linear programming solution might be unattainable at a reasonable cost.

The large-scale electronic computer became a reality in 1946 at the University of Pennsylvania, a year before Dantzig first developed and successfully applied linear programming. The first successful solution of a linear programming problem on a high-speed computer took place in January, 1952, on the National Bureau of Standards SEAC computer, and the following year computers became available on a production-line basis. Of course, today, almost every business and university has access to a computer. Most computers are now equipped with programs especially designed to solve problems of the linear programming type.

[1] W. W. Leontief, "Quantitative Input and Output Relations in the Economic System of the United States," *Review of Economics and Statistics*, August 1936.

[2] George B. Dantzig, "Maximization of a Linear Function of Variables Subject to Linear Inequalities," in T. C. Koopman's (ed.), *Activity Analysis of Production and Allocation* (New York: John Wiley, 1951).

THE SCIENTIFIC METHOD

Economics, engineering, and the sciences have had a long association with the scientific method. In contrast, business management's use of this method is a relatively recent development. This does not mean that prior to its use managerial problems were nonexistent; nor does it imply that management did not use the scientific method implicitly in their attempts to solve business problems. It simply recognizes that a systematic approach to the formulation and solution of business problems is an explicit characteristic of modern management.

There are numerous ways of defining the scientific method.[3] For our purposes, it embodies the following steps: (1) observation, (2) forming a hypothesis, (3) testing the hypothesis, and (4) repeating steps (1) through (3) if necessary.

Observation

The first step in the scientific method involves understanding the nature of the problem at hand. This means that before we can begin to do anything constructive we must "live with" the situation for a time. The significance of this becomes apparent when we realize that problem situations often contain a fundamental problem and a series of related problems. Although solving the related problems may improve the situation, the greatest benefit results when the fundamental problem is solved. Thus, we can say that the observation phase consists of acquiring an understanding of the situation so that we can define the fundamental problem correctly. At the same time, we can be gathering data concerning the possible causation variables for use in steps (2) and (3) of the scientific method.

Although the above is sufficiently general to define the observation phase, management is quite often interested in defining problems in a manner that will permit the formulation of optimal strategies. Optimal denotes "best possible." In one case optimal may refer to greatest profit, whereas in another case it may refer to least cost. Optimal-type problems require the choice of a criterion of effectiveness during the observation phase. This criterion enables one to evaluate a system's performance and also helps to identify the variables that exert influence on the achievement of optimality.

[3] E. H. Bowman and R. B. Fetter, *Analysis For Production Management* (Homewood, Ill.: Richard D. Irwin, Inc., 1961), Chapter 1; and T. L. Saaty, *Mathematical Methods of Operations Research* (New York: McGraw-Hill Book Co., 1959), Chapter 2.

The criterion of effectiveness should be consistent with some over-all objective. Consistency may require the use of suboptimization, since it is often difficult to relate explicitly the over-all objectives and the functioning of component parts. For example, a decision in the production department cannot be evaluated fully until all the ramifications of the decision are recognized and related to over-all company objectives. Realistically, this is not always possible; often the organizational components must suboptimize. That is, they must operate according to some proximate criteria that are only intuitively associated with over-all objectives; for example, production departments may utilize cost as their objective to be optimized.

Forming a Hypothesis

The second step in the scientific method is to develop a tentative hypothesis or explanation relating the dependent variable(s) and the causation variable(s). This step is alternatively called model building, and the resulting hypothesis is called a model. Model building is a process that relies heavily upon abstraction, whereby only the most important elements or features of a system are used to produce a useful explanation of the system. Abstracting from the real world is necessary in order to reduce significantly complex problems to manageable proportions. The end result of the process of abstraction is the model. Models are subject to various classifications, of which we shall use the following: physical models, schematic models, verbal models, and mathematical models. It is worthwhile to note that the process of abstraction or model formulation requires the use of simplifying assumptions. Any statement of the model should include a statement of the assumptions that have been made, lest characteristics be attributed to the model that it does not possess.

Physical models are representations of the real world without actually existing in the real world. A model ship used by naval architects to test hull designs in model basins is a good example of a physical model. In this case the important features of the real world that the model possesses are the contours of the hull. The model does not need engines, superstructure, or other appendages in order to produce useful results.

Schematics are line drawings, flow charts, graphs, and similar items that represent diagrammatically the major features of a particular system. Charts picturing the movement of work through a process need not show the details of the equipment included in the process in order to give industrial engineers a basis for recommending changes.[4]

[4] For more information on this topic see Bowman and Fetter, Chapter 3.

Verbal models are oral statements that define a system according to the measure of effectiveness, the causation variables, and their interrelationship. There are a number of deficiencies associated with verbal models that severely limit their usefulness. They are not precise enough to be formally tested, because language itself is vague and because complex situations are almost impossible to present verbally. Verbal models are useful in the initial stage of forming a hypothesis and usually lead directly to mathematical models.

The deficiencies of verbal models are substantially eliminated by the use of mathematical models. A simple mathematical model may appear as

$$S_t = aS_{t-1} - b(P_t - P_{t-1}).$$

That is to say, the sales of a firm in time period t are a function, other things being equal, of the sales made in period $t - 1$, the price charged in period t, and the price in period $t - 1$. The letters a and b are parameters that give weight to the causation variables, sales in $t - 1$ and price. This model of sales activity may be used with a cost model to predict the profits of the firm for time period t. A simple cost model might appear as the following:

$$TC_t = a + bx.$$

That is, the total cost of operating in period t is a function of a fixed amount a (fixed cost) and a variable amount bx (variable cost), where b is a parameter and x denotes the volume of output in the period t. The assumptions of this model are too lengthy to present here, but for most firms these assumptions are highly restrictive.[5] The completed model, which gives the profit for the firm in time period t, consists of the above equations for sales and costs and the equation for profit: $\pi_t = S_t - TC_t$— which means that profit is the difference between sales and total costs for the period. This last equation is the measure of effectiveness or objective function.

Testing the Hypothesis

Once the model has been constructed, it must be tested. Up to this point the model is merely a tentative explanation. Any attempt to use the model predictively, in the operating sense, is fraught with danger. Basically, models may be tested by

[5] M. Spencer and H. Spiegelman, *Managerial Economics* (Homewood, Ill.: Richard D. Irwin, 1959), pp. 114–116.

either of two methods. The first approach is to see how well the model will explain what has happened. The second is to find out how well the model is capable of predicting what will happen. Because the former method is somewhat biased by the use of past information in the formulation of the model, the latter method is preferable. To illustrate the testing procedure we shall refer to our mathematical model involving the determination of profit. By reference to the records of the company one could determine whether or not the model sufficiently explained the past. At the same time one could compare future prices and information with past sales and prices (of period $t - 1$) to predict the profits for period t. If the model does not predict with sufficient accuracy to be useful, it must be modified. This requires that the researcher repeat steps (1) through (3) of the scientific method. Repetition of these steps may reveal that advertising expenditures should be included in the sales function. The improved model of sales activity may then appear as follows:

$$S_t = aS_{t-1} - b(P_t - P_{t-1}) - c(A_t - A_{t-1}).$$

Testing the improved sales activity model may show that the results are sufficiently accurate to be useful even though they still may not be exact. The usefulness of a mathematical model and models in general is easily recognized by reference to our simple example. If the firm wishes to know the effect of a change in the price of the product on the level of profit, it merely manipulates the mathematical model. This is obviously to be preferred to the alternative of changing the price of the product and waiting for the market to produce the results. Similarly, the firm might want to investigate the effect on profit of an increase in fixed cost. This can be accomplished by changing the parameter a in the model for cost. The latter example brings to light our final point regarding models. The fact that a useful model has been successfully constructed does not mean that the model can be used blindly from that point forward. The model must be checked continuously against new developments in the system. Changes may be taking place that will cause the functional relationships to be altered. Thus, model building is a continuing process and the scientific method provides the approach necessary for successful model construction.

MANAGEMENT AND LINEAR PROGRAMMING

A basic function of management is decision making—that is, the process of choosing from among a number of alternative courses of action. Decisions are made concerning the objectives of the firm: what, how much, and when to produce;

the nature of personnel policies; and other questions. The list is endless. In order to make these decisions management requires information. In most situations there is an enormous amount of information available to management, and the problem is to choose the relevant information and process it as quickly as possible. Thus it is that management often makes decisions in spite of, rather than in light of, the information available. To clarify this point, let us consider the following simple problem.

A single-plant firm produces a number of fairly similar products. Each product requires a number of machining operations, which are performed on a number of different machines. In addition, each machine requires varying lengths of time to perform any one operation. The length of time required to produce any particular product with the same set of machines is not consistent, nor are the raw material requirements for each product exactly equal. Let us assume that management is interested in finding the particular combination of products that should be produced. There is information available concerning the process time per product per operation per machine, raw material availabilities in terms of both quantities and prices, hours of machine availability, demand prices and quantities, plus a host of other factors that may or may not have some effect on the extent to which a solution is achieved. Clearly the number of combinations that can be produced may be very large. In fact, there may be so many technically feasible combinations that it would be very difficult to decide which combinations are best. It is little wonder, then, that many management problems are solved by rules of thumb that generally tend to produce less than optimal results.

Faced with this type of problem, which belongs to a larger class of problems called allocation problems, management can turn to linear programming for assistance. Linear programming involves the use of a mathematical model to provide the optimal solution to certain types of problems having the following characteristics:

1. There is an objective function that management has decided will most effectively measure the "goodness," or optimality, of any proposed solution. This measure of effectiveness may be stated in terms of the particular product mix that maximizes the contribution to profit and overhead. Alternatively, the measure of effectiveness could be stated in terms of the particular product mix that minimizes machining time.

2. The restraints that impose limits on variables, which in turn limit the extent to which the objective function can be optimized (maximized or minimized), must be known. In terms of our illustrative problem,

the extent to which the contribution to profit and overhead could be maximized is limited, among other things, by the hours of machine availability, raw materials availability, and processing times.

3. The objective function and solution restraints can be expressed algebraically. When these three conditions are satisfied, the result is a mathematical model describing the activity under study.
4. In order for the model developed in steps 1 through 3 above to be a linear programming model, the relationships included in the model must be linear.

LINEARITY

A linear function is expressed in the following form:

$$F(X_1, X_2, \ldots, X_n) = a_1X_1 + a_2X_2 + \cdots + a_iX_i + \cdots + a_nX_n,$$

or

$$F(X_1, X_2, \ldots, X_n) = \sum_{i=1}^{n} a_iX_i,$$

where $\sum_{i=1}^{n} a_iX_i$ is simply mathematical shorthand for $a_1X_1 + a_2X_2 + \cdots + a_nX_n$. The following is an example of a typical linear function:

$$F = 2X_1 + 4X_2 + 1X_3 + 5X_4.$$

A linear function may be expressed as a straight line on a graph, if the function contains only two variables. The linear function $Y = mX + b$ may be expressed graphically as

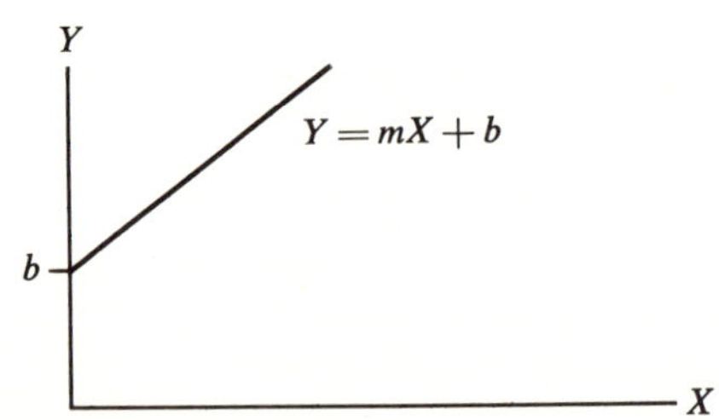

where m is the slope of the line and b is the Y-intercept. This is known as the slope-intercept form of the graph of a straight line. For example, the equation

$$4X_1 + 8X_2 = 10$$

may be placed in slope-intercept form by solving for X_1,

$$X_1 = -2X_2 + 2.5,$$

where -2 is the slope of the line and 2.5 is the value of the intercept on the X_1 axis.

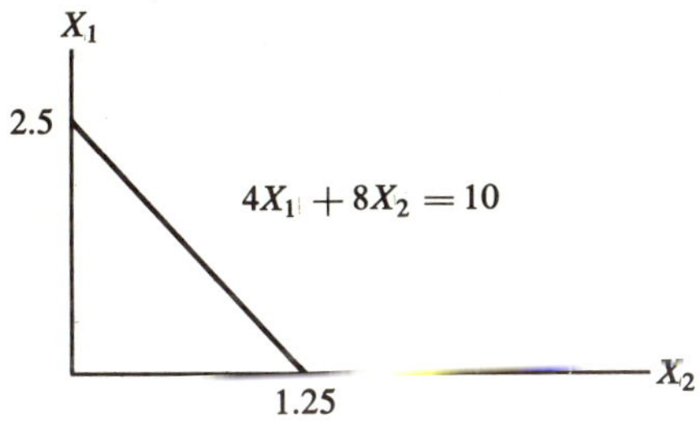

The slope-intercept form will be useful in the solution of problems by the graphical method presented in Chapter 3.

To summarize, linear programming is a mathematical technique for solving certain types of managerial decision problems whose objective or measure of effectiveness can be expressed as a linear function of a set of independent variables subject to a set of linear restraints. The first two conditions above provide the clues for determining the ways in which management can benefit from the use of linear programming.

BENEFITS DERIVED FROM THE USE OF LINEAR PROGRAMMING

We have already indicated that linear programming will provide the optimum solution to a problem when certain conditions exist. In addition, linear programming:

1. *Aids Understanding.* When management makes the decision to use linear programming, it is committing itself to a new way of thinking about business problems. Because of the mathematical nature of the technique, vague generalizations describing the firm's activities are no longer adequate. Management must thoroughly understand the firm's activities before it can attempt to represent them mathematically. This requires that the relationships governing the performance of an activity be fully understood and, in addition, the relationship of any given activity to the over-all performance of the firm must be known as well. The result of quantifying such relationships is the development of a greater understanding of the firm's activities.

2. *Provides Information for Planning and Control.* The use of linear programming further contributes to the fulfillment of the managerial function by providing information that is useful in planning and control. In part, the planning function is concerned with the recognition of change. This aspect of planning is aided by the susceptibility of the linear programming model to manipulation.[6] Thus we have questions such as, what is the effect on the present mix of products if certain raw materials' prices change? Or, what is the effect of a prolonged machine breakdown, or an increase in machinery downtime? Control over operations is achieved by comparing plans with results. To the extent that linear programming is useful in planning it will also contribute to the fulfillment of the managerial function of control.

3. *Contributes to the Development of Executives.* The matter of executive development is a serious problem in many firms. Training programs are valuable in this respect, and their value can be increased through the use of linear programming and other mathematical models as training devices. Future executives can be given actual or fictitious problem situations in which they are required to develop the mathematical model. Their results can easily be verified without recourse to actual operations where mistakes are likely to prove quite costly. Much of the knowledge and confidence that ordinarily takes many years of experience to obtain can be gained in a significantly shorter period of time via the technique of model-building.

It should be obvious that all of the above benefits will create a more efficient management, which will result in increased profitability for the firm.

Before we become completely entranced by the potential benefits of linear programming we should be made aware of its limitations. It cannot eliminate all of the judgment that is required for managerial decison making. Although linear programming may eliminate managerial judgment in such decision areas as product mix, order allocation, machine scheduling, and warehouse location, management must still decide, for example, whether the best measure of effectiveness for a particular activity is the contribution to profit and overhead, or whether it is best to minimize processing time. Likewise, it must make decisions on the effects of certain programs on employee morale and decide whether or not certain products must be produced so as to fill out the product line. These are elements of decision making that management retains, and linear programming does not diminish their importance.

[6] This point is referred to as sensitivity analysis, or parameter programming. See Chapter 6.

LINEAR PROGRAMMING APPLICATIONS

Linear programming has been used with problems of optimizing executive compensation programs, location of and management of river dam projects, farm management, awarding contracts, traffic control, scheduling a military tanker fleet, minimizing trim losses in paper mills, balancing assembly-line operations, and optimizing investment portfolio allocations for both individual and institutional investors.

The agricultural, industrial, and military sectors have made the most extensive use of linear programming. In addition, many applications can be found in the sciences, including economics and applied engineering. There is little doubt that the list of applications will continue to grow as decision makers become more familiar with the potentials of the method. Concurrently, the continuing development of high-speed computing equipment will be a major factor in the application of linear programming to those problems that fit into the framework of the technique.

In this section a few of the broader applications are presented, with the hope that they will sufficiently indicate the general problem types that linear programming can solve. The following examples are typical of the general problem areas in which linear programming has been employed successfully.[7]

Scheduling Production and Inventory Control

Many firms—for example, oil refining companies—produce products that are subject to seasonal sales fluctuations. In such cases, widely fluctuating production rates have proved extremely costly, whereas a uniform production rate builds up inventories that result in excessive storage costs. Through the use of linear programming, these companies can establish a production schedule that will satisfy demand when it arises as well as minimize both the costly effects of fluctuating production rates and the excessive inventory carrying charges resulting from a uniform production rate.

Further applications in this area involve finding the best uses of production capacity and the best basis for assigning production to a number of manufacturing or producing centers. Excess capacity problems involve selecting from among a group of

[7] For a comprehensive bibliography of linear programming applications see V. Riley and S. Gass, *Linear Programming and Associated Techniques* (Baltimore: The Johns Hopkins Press, 1958).

potential products that set of products which will both utilize the capacity and maximize the contribution to profit and overhead. Assigning production to a number of producing points with due consideration to the initial location of raw materials, plant production capacities, demand requirements, and final destination of products is a problem type that is easily solved by linear programming. This is also similar to the task of assigning production to processes in a plant where the products have varying production periods and the processes are not equally efficient. In this instance linear programming will provide for the optimum allocation of products among the different processes in accordance with some objective, such as minimum processing time.

Blending Problems

There are numerous instances in which certain basic components are combined to produce a product that has a certain set of specifications. Examples of this type of problem are to be found in the blending of gasolines, the mixing of cattle feeds, and the mixing of meats to produce sausage or other meat products. A problem in oil refineries, for example, is to select a number of crude oil products to be blended into different grades of gasoline. The basic products have certain common characteristics, such as octane rating, vapor pressure indices, and distillation temperatures. Considering the demand for gasoline of varying specifications and the supply conditions for the crude oil products, linear programming can determine the mixture of crude products that will yield maximum profits, minimum costs, or some other optimum. The cattle-feed problem concerns minimizing the cost of producing a feed that possesses certain nutritional requirements.

Purchasing

Processes requiring inputs that are available at different quantities, qualities, and prices present a purchasing problem that can be solved for a least-cost objective. An oil refinery, which purchases all of its crude oil from outside sources, uses linear programming with a profit maximizing objective to determine which crudes should be purchased. The model considers output requirements and specifications, crude oils on hand, and processing costs in the evaluation of this problem.

Whether to make a product component or purchase it from an outside source is another problem area in which linear programming can be of assistance. The factors that are considered in a problem of this type are: process and product requirements, production and purchasing costs, selling costs, overhead costs, and either a profit or cost objective to be optimized.

Routing and Assignment

The route that a salesman should take in order to cover a specified territory while minimizing distance traveled is also a problem type that can be solved by linear programming. Under certain conditions the transportation method can be used for any routing problem involving an origin-destination scheme. Thus, we may have empty boxcars at various locations (origins) and requests for boxcars at various locations (destinations). Given the costs of moving empty boxcars, the transportation technique will give the optimum assignment of empty boxcars from the origins to the destinations. This technique is also useful in determining the location of additional warehousing facilities and the allocation of products to machines.

Another class of problems involves assigning facilities to jobs in a manner that embodies some performance objective. Thus, problems of assigning products to facilities in which the costs of producing each product will vary with the facility chosen can be solved by linear programming. This technique can also be used to assign workers to jobs offering a quantitative measure of the workers' effectiveness.

2 A Survey of Linear Programming Methods

Up to now we have examined, in a general manner, the scientific approach to management with special emphasis on mathematical models and their usefulness in solving managerial decision problems. We have also discussed the general nature of a specific mathematical model, linear programming, and its relationship to the managerial decision-making process. Finally, we have indicated the general types of problems that can be solved by this technique.

Having established the importance of linear programming as an individual tool of management, we shall now investigate some of the more important concepts and techniques used in solving linear programming problems. Specifically, this chapter is concerned with:

1. A nonmathematical summary of four different computational procedures for solving linear programming problems.

2. A mathematical statement of the types of problems that can be solved by each of these four methods. (These statements are preceded by an asterisk to indicate that this information is not a prerequisite to understanding how to apply the particular method to practical problems.

Furthermore, these mathematical statements presuppose some degree of familiarity with mathematical notation.)

3. A brief description of four special techniques: linear programming under uncertainty, nonlinear programming, integer programming, and dynamic programming. (These programming techniques are designed to solve problems that are more complex than those which can be solved by straightforward linear programming methods.)

Before we begin our survey we should first recall that linear programming is a mathematical technique that will give the optimum solution to problems defined by a linear objective function subject to a set of linear restraints. The solution is called a program. Mathematically, the problem of linear programming is one of optimizing (maximizing or minimizing) an objective function of the following form:

$$Z = \sum_{j=1}^{n} C_j X_j \qquad (j = 1, 2, \ldots, n),$$

subject to restraints of the form

$$\sum_{j=1}^{n} a_{ij} X_j \begin{pmatrix} \leq \\ = \\ \geq \end{pmatrix} b_i \qquad (i = 1, 2, \ldots, m),$$

and

$$X_j \geq 0 \qquad (j = 1, 2, \ldots, n),$$

where

X_j = the quantity of the jth variable of interest to the decision-maker, where there are n variables being considered;

C_j = the per unit contribution to the objective function (profit or cost) of the jth variable, where there are n variables;

Z = the objective function to be maximized or minimized;

a_{ij} = the exchange coefficient of the jth variable in the ith restraint, where there are m restraints and n variables;

b_i = the ith requirement, where there are m requirements in all.

The linear restraints may take one of three possible forms:

1. less than or equal to ($\leq$),
2. equal to ($=$),
3. greater than or equal to ($\geq$).

Restraints expressed as ($\leq$) indicate that the value of the left side of the restraint expression cannot exceed the value of the right side of the expression. Restraints expressed as ($=$) require that the value of the left side of the restraint expression be exactly equal to the value of the right side of the expression. Restraints expressed as ($\geq$) require that the value of the left side of the restraint expression be at least as large as that of the right side of the expression. The expression ($X_j \geq 0$) simply means that the X_j's must be non-negative.

The mathematical notation used above conforms to that which is most widely used in the literature. Similar notation is used in the mathematical formulation of the graphical, simplex, transportation, and assignment problems.

The following is the mathematical statement of a typical linear programming problem.

Minimize:

$$Z = 10X_1 + 8X_2 + 12X_3;$$

subject to:

$$\begin{aligned} 6X_1 + 9X_2 + 7X_3 &\geq 6000, \\ 2X_1 + 6X_2 + 3X_3 &\geq 3000, \\ 1X_1 + 5X_2 + 2X_3 &\leq 4000, \\ X_1 + X_2 + X_3 &= 1500; \end{aligned}$$

where:

$$X_1, X_2, X_3 \geq 0.$$

THE GRAPHICAL SOLUTION

Certain linear programming problems can be solved by graphical analysis. The limitations of this technique are the impossibility of drawing graphs in more than three dimensions, the amount of time required to prepare the graphs, and the relative imprecision of graphical analysis. Nevertheless, the graphical method is useful as an expositional device, and is the subject of Chapter 3. Briefly, graphical solutions to linear programming problems involve:

1. plotting the restraints on a graph,
2. plotting the objective function on the same graph, and
3. using 1 and 2 to determine the solution that satisfies the constraints and optimizes the objective function.

* *Mathematical Statement—Graphical Problem*

1. Objective Function To Be Optimized:

$$Z = \sum_{j=1}^{3} C_j X_j \qquad (j = 1, 2, 3).$$

2. Subject to Restraints of the Form

$$\sum_{j=1}^{3} a_{ij} X_j \begin{pmatrix} \leq \\ = \\ \geq \end{pmatrix} b_i \qquad (i = 1, 2, \ldots, m),$$

and

$$X_j \geq 0 \qquad (j = 1, 2, 3).$$

The limitation on j $(j = 1, 2, 3)$ is derived from the fact that it is impossible to draw graphs in more than three dimensions. In practice, three-dimensional graphs are difficult, if not impossible, to work with, and graphical analysis usually is restricted to the two-dimensional case.

THE SIMPLEX METHOD

Although a number of computational techniques can be used for special types of linear programming problems, the simplex procedure is the general method. It can be used to solve any linear programming problem, and its procedure will be presented in detail in Chapter 10. The following is a verbal description of the simplex procedure.

Given the conditions of a linear objective function and a set of linear restraints, an initial solution is developed. The simplex procedure will determine whether or not the solution can be improved by changing the program. If improvement is possible, a new program will be developed. It will also be tested to see if it is the optimum program and, if it is not, further program changes will be made. This program evaluation procedure will continue until it is no longer possible to improve the solution further, indicating that an optimum solution has been found. Summarily, the simplex method is an iterative technique that successively improves the solution until an optimum solution is found.

The fact that many linear programming problems require the use of an electronic computer has led to the development of what is called the Revised Simplex Method. This method has a number of computational features that make it particularly suitable for solving problems on an electronic computer. The basic difference between the revised simplex and the original simplex rests in the number of calculations necessary to move from one iteration to the next. The revised simplex method may require fewer calculations per iteration than the simplex method. Furthermore, the tableaux used with the revised simplex method need not contain as many entries as the tableaux used in the original simplex method, thus simplifying recording problems which use valuable computer memory space. We shall concentrate on the presentation of the original simplex procedure rather than the revised framework. The revised simplex method is mentioned at this point because it is the method used most often when simplex codes are written for computers.[1]

* *Mathematical Statement—Simplex Problem*

1. Objective Function To Be Optimized:

$$Z = \sum_{j=1}^{n} C_j X_j.$$

2. Subject to Restraints of the Form

$$\sum_{j=1}^{n} a_{ij} X_j \begin{pmatrix} \leq \\ = \\ \geq \end{pmatrix} b_i \qquad (i = 1, 2, \ldots, m),$$

and

$$X_j \geq 0 \qquad (j = 1, 2, \ldots, n).$$

[1] Those readers who are interested in actually writing linear programming computer codes using the Revised Simplex Method may see Saul I. Gass, *Linear Programming* (New York: McGraw-Hill Book Co., 1958), Chapter 6. However, a complete understanding of the Revised Simplex Method requires considerable knowledge of mathematics.

THE TRANSPORTATION AND ASSIGNMENT METHODS

We have noted in the section above that the simplex method can be used to solve any linear programming problem. However, for certain types of problems there are available computational procedures that reduce the complexity of the simplex procedure. Two of these techniques are the transportation method and the assignment method. Both of these methods, discussed briefly in this section, are treated in much greater detail in Chapter 7.

The Transportation Method

This particular method was developed to handle a special class of linear programming problems in which the sum of resources is equal to (or can be made equal to) the sum of the requirements. For example, a firm may have three factories and four warehouses to which products are shipped. The cost of shipping from each factory to each warehouse is different, whereas total factory output (resources) is equal to warehouse needs (requirements). Such a problem may be represented in the form of a matrix with the unit shipping costs included as elements of the matrix.

TABLE 2-1

	Warehouse				
Factory	W_1	W_2	W_3	W_4	*Factory Capacities*
F_1	\$0.30	\$0.25	\$0.40	\$0.20	100
F_2	\$0.29	\$0.26	\$0.35	\$0.40	250
F_3	\$0.31	\$0.33	\$0.37	\$0.30	150
Warehouse Requirements	100	150	200	50	500

The transportation method provides a shipping schedule that minimizes total transportation costs. The transportation technique is an iterative procedure that develops an initial solution and determines whether or not changes in the program will improve the solution. Changes are made in accordance with a computational technique that successively improves the original solution until the optimum solution is found. The usefulness of the method is not limited to those cases in which physical quantities are shipped from sources to destinations. It is limited to those cases in which (1) total resources equal total requirements, both of which are stated in only one kind of unit; and (2) all of the coefficients of the linear restraints are either zero or one.

* *Mathematical Statement—Transportation Problem*

1. Objective Function To Be ptimized:

$$Z = \sum_{i=1}^{m} \sum_{j=1}^{n} C_{ij}X_{ij} \qquad (i = 1, 2, \ldots, m;\; j = 1, 2, \ldots, n).$$

2. Subject to Restraints of the Form

$$X_{ij} \geq 0 \qquad (i = 1, 2, \ldots, m;\; j = 1, 2, \ldots, n),$$

and

$$\sum_{i=1}^{m} X_{ij} = a_j \qquad (j = 1, 2, \ldots, n),$$

$$\sum_{j=1}^{n} X_{ij} = b_i \qquad (i = 1, 2, \ldots, m),$$

$$\sum_{j=1}^{n} a_j = \sum_{i=1}^{m} b_i.$$

The mathematical notation used above may best be interpreted by relating it to the sample problem involving the scheduling of factory output to warehouses.

X_{ij} = the number of product units shipped from the ith factory to the jth warehouse;

C_{ij} = the cost of shipping one product unit from the ith factory to the jth warehouse;

m = 3 (the number of factories);

n = 4 (the number of warehouses);

a_j = the requirement of the jth warehouse;

b_i = the capacity of the ith factory.

The objective of the firm is to minimize total transportation costs for all possible factory-warehouse shipping patterns. (There are 3 × 4 = 12 possible shipping combinations.) In other words, the firm wants to minimize

$$Z = \sum_{i=1}^{3} \sum_{j=1}^{4} C_{ij}X_{ij} \qquad (i = 1, 2, 3;\ j = 1, 2, 3, 4),$$

or

$$\begin{aligned} Z = {} & 0.30X_{11} + 0.25X_{12} + 0.40X_{13} + 0.20X_{14} \\ & + 0.29X_{21} + 0.26X_{22} + 0.35X_{23} + 0.40X_{24} \\ & + 0.31X_{31} + 0.33X_{32} + 0.37X_{33} + 0.30X_{34}, \end{aligned}$$

subject to the following restraints:

1. *Requirements* (Demand)

$$\sum_{i=1}^{3} X_{ij} = a_j \qquad (j = 1, 2, 3, 4);$$

$$\begin{aligned} X_{11} + X_{21} + X_{31} &= 100, \\ X_{12} + X_{22} + X_{32} &= 150, \\ X_{13} + X_{23} + X_{33} &= 200, \\ X_{14} + X_{24} + X_{34} &= 50. \end{aligned}$$

2. *Capacity* (Supply)

$$\sum_{j=1}^{4} X_{ij} = b_i \qquad (i = 1, 2, 3);$$

$$\begin{aligned} X_{11} + X_{12} + X_{13} + X_{14} &= 100, \\ X_{21} + X_{22} + X_{23} + X_{24} &= 250, \\ X_{31} + X_{32} + X_{33} + X_{34} &= 150. \end{aligned}$$

3. *Demand = Supply*

$$\sum_{j=1}^{4} a_j = \sum_{i=1}^{3} b_i;$$

$$100 + 150 + 200 + 50 = 100 + 250 + 150,$$

$$500 = 500.$$

This problem can be solved by either the simplex method, or the transportation method. However, the transportation method should prove to be the more efficient alternative in terms of computational time.

The Assignment Method

The second special technique we will discuss is the assignment method. It too offers computational efficiency over the simplex method when certain conditions are satisfied. Suppose that there are 4 men available to perform 4 jobs, and the performance characteristics of each man for each job are known. The assignment method of linear programming will assign the 4 men to the 4 jobs in a manner which optimizes their performance characteristics. This problem may be represented in matrix form with the performance characteristics included as elements.

TABLE 2-2

	Job 1	*Job* 2	*Job* 3	*Job* 4
Man 1	10	8	11	7
Man 2	5	15	7	9
Man 3	8	12	10	7
Man 4	12	9	11	8

The performance characteristics may be production time, assignment costs, or profit. The assignment method utilizes a technique known as the reduction theorem to develop the optimal solution to a given problem.

* *Mathematical Statement—Assignment Problem*

1. Objective Function To Be Optimized:

$$Z = \sum_{i=1}^{n} \sum_{j=1}^{n} C_{ij} X_{ij} \qquad (i = 1, 2, \ldots, n;\ j = 1, 2, \ldots, n).$$

2. Subject to Restraints of the Form

$$X_{ij} = 0 \text{ or } 1 \qquad (i = 1, 2, \ldots, n;\ j = 1, 2, \ldots, n),$$

$$\sum_{i=1}^{n} X_{ij} = 1 \qquad (j = 1, 2, \ldots, n),$$

$$\sum_{j=1}^{n} X_{ij} = 1 \qquad (i = 1, 2, \ldots, n).$$

This mathematical notation can be used to formulate the relatively simple problem of assigning 4 men to 4 jobs, as described above.

$X_{ij} = 1$ (if the ith man is assigned to the jth job);

$X_{ij} = 0$ (if the ith man is *not* assigned to the jth job);

$C_{ij} =$ the performance characteristic of assigning the ith man to the jth job;

$n = 4$ (the number of men and jobs).

The objective is to optimize the total performance characteristic of assigning 4 men to 4 jobs. In other words, we want to maximize or minimize Z (depending on the nature of the performance characteristic), where

$$Z = \sum_{i=1}^{4} \sum_{j=1}^{4} C_{ij} X_{ij} \qquad (i = 1, 2, 3, 4;\ j = 1, 2, 3, 4),$$

or

$$\begin{aligned} Z = X_{11} &+ X_{12} + X_{13} + X_{14} \\ &+ X_{21} + X_{22} + X_{23} + X_{24} \\ &+ X_{31} + X_{32} + X_{33} + X_{34} \\ &+ X_{41} + X_{42} + X_{43} + X_{44}, \end{aligned}$$

subject to the following restraints:

1. $X_{ij} = 0$ or $1 \qquad (i = 1, 2, 3, 4;\ j = 1, 2, 3, 4).$

2. Each job must be assigned to one and only one man.

$$\sum_{i=1}^{4} X_{ij} = 1 \qquad (j = 1, 2, 3, 4);$$

$$X_{11} + X_{21} + X_{31} + X_{41} = 1,$$
$$X_{12} + X_{22} + X_{32} + X_{42} = 1,$$
$$X_{13} + X_{23} + X_{33} + X_{43} = 1,$$
$$X_{14} + X_{24} + X_{34} + X_{44} = 1.$$

3. Each man must be assigned to one and only one job.

$$\sum_{j=1}^{4} X_{ij} = 1 \qquad (i = 1, 2, 3, 4);$$

$$X_{11} + X_{12} + X_{13} + X_{14} = 1,$$
$$X_{21} + X_{22} + X_{23} + X_{24} = 1,$$
$$X_{31} + X_{32} + X_{33} + X_{34} = 1,$$
$$X_{41} + X_{42} + X_{43} + X_{44} = 1.$$

This problem may be solved by the simplex method or the transportation method, but solution by the assignment method will be the most efficient method, if one wishes to minimize computation time.

SPECIAL TOPICS

Throughout our discussion of linear programming we have made three assumptions:

1. Linearity,
2. Complete certainty, and
3. Stationary or static time conditions.

However, in attempting to solve many of the problems of the real world these assumptions are completely invalid. One would be making a very serious mistake if he were to attempt forcing a linear programming solution on to a problem in which one or more of these assumptions is violated. However, computational procedures have been developed for solving problems involving nonlinearity, uncertainty, and dynamic time considerations. Unfortunately, these computational procedures are considerably more complex than those used for solving linear programming problems. Hence, an extended treatment of these procedures is beyond the scope of this text, and only a brief sketch of these procedures will be presented here. The objective of this section is simply to provide some guidance as to where to look for additional information in the event that a particular problem situation does not conform to the rigid assumptions of the linear programming model.

For certain types of problems it is often necessary or desirable to express the solutions as whole numbers (integers). However, since the simplex method does not always provide integral solutions to linear programming problems, we have included a brief discussion of integer programming.

Linear Programming under Uncertainty

Until recently, linear programming was limited in its application to problems involving decision-making under complete certainty, or at least to problems whose solutions could be approximated by an assumption of complete certainty, such as deterministic problems. Decision making under complete certainty exists when each alternative course of action available to the decision maker leads to one and only one final outcome. However, the recent mathematical contributions of George Dantzig and others have made it possible to solve linear programming problems under conditions of uncertainty, such as probabilistic problems. Decision making under uncertainty refers to the class of problems in which a set of possible final outcomes is associated with each alternative course of action available to the decision maker, and the probability of the occurrence of each of these outcomes is known. Linear programming problems in which either the technological coefficients or the coefficients of an objective function are random variables are known as stochastic linear programming problems.[2]

[2] Readers interested in a comprehensive survey of stochastic linear programming or linear programming under uncertainty should see A. Madansky's "Linear Programming under Uncertainty," in R. Graves and P. Wolfe, ed., *Recent Advances in Mathematical Programming* (New York: McGraw-Hill Book Co., 1963).

Nonlinear Programming[3]

Although linearity is a necessary assumption for the formulation and solution of all of the problems included in this book, it is only natural to consider problems that do not fit into the framework of the linear programming technique, because either the objective function or the restraints cannot realistically be formulated in linear terms. Most of the work that has been done in the area of nonlinear programming falls into one of three categories: (1) approximation methods; (2) quadratic methods; (3) Lagrangian methods. Since the solution of nonlinear programming problems by any of these methods requires considerable knowledge of analytical geometry, calculus, and the theory of equations, the discussion of these techniques will necessarily be limited to considerations of a general nature.

Approximation Methods. In the case of a nonlinear objective function it may be possible to obtain a solution to a problem by using a set of linear approximations to the objective function, each of them valid within a given interval. For example, the following curvilinear function, $Z = f(X)$, might be approximated by

$$Z = f_1(X_1) + f_2(X_2) + f_3(X_3)$$

(see Figure 2-1), where $f_1(X_1)$, $f_2(X_2)$, and $f_3(X_3)$ are linear functions of X_1, X_2, and X_3 respectively, and where

$$a_0 \leq X_1 < a_1,$$
$$a_1 \leq X_2 < a_2,$$
$$a_2 \leq X_3 \leq a_3.$$

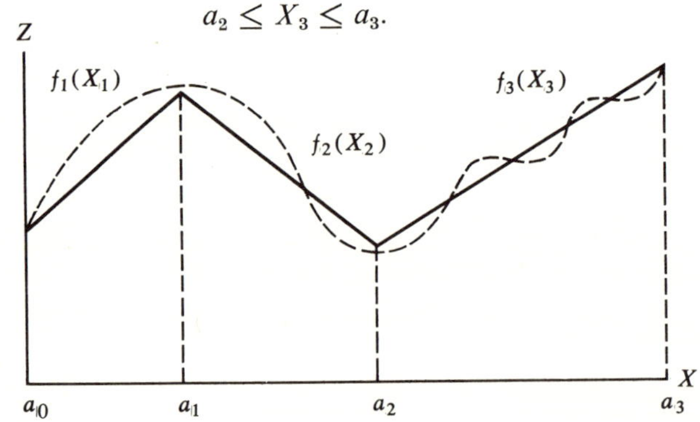

Figure 2-1.

[3] For an introductory treatment, see T. H. Naylor and J. M. Vernon, *Microeconomics and Decision Models of the Firm* (New York: Harcourt, Brace and World, 1969), Chapters 6 and 9.

Once a nonlinear objective function has been expressed as the sum of a set of linear functions, the simplex method can be used to find an optimum solution to the problem. The more equations that are used to approximate the objective function, the more accurate the final solution will be, but this advantage may be offset by the time required to solve these equations. Linear approximation techniques have been developed to handle nonlinearity in both objective functions and in restraints involving any number of variables.

Quadratic Methods. A technique known as quadratic programming has been developed for finding the minimum value of a quadratic function subject to linear restraints. (A quadratic function is a function that consists of terms of second degree as the highest.) For example, quadratic programming can be used to determine the non-negative values of X_1 and X_2 that will minimize

$$Z = X_1^2 + 3X_2^2 - 4X_1 - 6X_2,$$

subject to

$$X_1 + 2X_2 \leq 4.$$

With the aid of quadratic programming, solutions may be obtained for economic problems concerning the maximization of profit for the firm when production functions are linear and marginal costs vary linearly with output. Quadratic approximation may also be used to minimize general nonlinear objective functions subject to linear restraints.

Lagrangian Methods. The Lagrangian multiplier technique utilizes the calculus of maxima and minima to maximize or minimize a function of $(X_1, X_2, \ldots, X_n)$ subject to $f_j(X_1, X_2, \ldots, X_n) = 0$; $(j = 1, \ldots, m)$. Both the objective function and the restraints may be nonlinear in problems solved by this method. Two major limitations on the unmodified version of the Lagrangian technique are (1) negative solutions may arise and (2) the restraints must be in the form of equalities rather than inequalities. However, H. W. Kuhn and A. W. Tucker have developed an extension to this method that guarantees non-negative solutions and permits restraints that are expressed as inequalities.[4]

[4] H. W. Kuhn and A. W. Tucker, "Nonlinear Programming," *Proceedings of the Second Berkeley Symposium on Mathematical Statistics and Probability*, J. Neyman, ed. (Berkeley, Calif.: University of California Press, 1951), pp. 481–492.

Integer Programming

Certain types of linear programming problems are of such a nature that their solutions must be expressed as integers, i.e., whole numbers such as 1, 4, 7, 500, etc., rather than as decimals or fractions, such as 1.78, $9\frac{1}{2}$, 748.998, etc. For example, a linear programming solution that proposes that $\frac{1}{4}$ of an aircraft be assigned to a route is of questionable value. Likewise, it is difficult to build $\frac{7}{8}$ of an electrical generating station or serve a patient in a hospital $\frac{75}{87}$ of a glass of orange juice. Each of these problems is characterized by some form of indivisibility.

Since the simplex method does not guarantee integral solutions, a method has been developed by R. E. Gomory for solving problems whose solutions must be in the form of integers.[5] This method starts with the optimum solution of a linear programming problem excluding the integral restraints. Gomory then applies his "cutting plane" method to the nonintegral optimum solution, and through an iterative process finally arrives at an optimum solution in terms of integers. Unfortunately, the Gomory "cutting plane" method is not an efficient computational method. This will be illustrated in Chapter 7 where we apply the Gomory method to a simple example.

Another, and rather different method for solving integer programming problems, is the "branch and bound" method. An exposition is contained in the book by Hillier and Lieberman.[6]

Dynamic Programming

Dynamic programming refers to a mathematical method of analysis that is used to solve multi-stage decision problems in which decisions generated by Stage 1 become conditions for Stage 2, etc. Production scheduling is an example of a multi-stage decision problem; the amount a firm plans to produce this month is partially dependent on how much they produced last month, because of the possible accumulation or reduction of backlog.

The essential characteristic of dynamic programming is that optimum solutions are reached in a step-by-step manner. Given the conditions of Stage 1, an optimum solution is determined. Then the optimum solution from Stage 1 is integrated with

[5] R. E. Gomory, "An Algorithm for Integer Solutions to Linear Programs," *Princeton–I.B.M. Mathematics Research Project Technical Report*, No. 1 (September 17, 1958).

[6] F. S. Hillier and G. J. Lieberman, *Introduction to Operations Research* (San Francisco: Holden-Day, 1967).

Stage 2 in order to arrive at a new optimum solution. The optimum solution of Stage 2 is then integrated with Stage 3 in order to arrive at a new optimum solution. This procedure is continued throughout all stages.

Mathematical techniques have been developed to handle lengthy multi-stage decision problems involving nonlinearity and uncertainty, as well as problems involving linearity and complete certainty. Frequently it may prove advantageous to use an electronic computer to solve problems of exceptional length. By far the most complete treatment available on dynamic programming is Richard Bellman's *Dynamic Programming*.[7]

[7] Richard Bellman, *Dynamic Programming* (Princeton, N.J.: Princeton University Press, 1957).

3 The Graphical Solution

In order to illustrate the graphical approach to problems that can be solved by linear programming, let us consider the following problem involving the manufacture of television receivers. Admittedly, this problem is quite simple and could be solved by either trial-and-error methods or marginal analysis. However, the knowledge acquired from solving this relatively simple problem by the graphical method and the simplex method should enable one to solve problems of a general and more complex nature—problems that are beyond the scope of trial-and-error and marginal analysis methods.

THE TELEVISION ASSEMBLY PROBLEM

A medium-scale manufacturer of television receivers has production facilities for assembling two different types of television sets. These facilities can be used to assemble both black-and-white and color sets. At the present time the firm is producing only one model of each type of set. The black-and-white set retails for \$198 and contributes \$15 to profit and overhead, whereas the color set sells for \$499.95 at the retail level and contributes \$45 to profit and overhead. Both the black-and-white sets and the color sets are 21-inch table models.

The television manufacturer is concerned with the problem of deciding how to best utilize his production facilities so as to maximize profits. Since labor skills and plant equipment are completely interchangeable with regard to the production of the two types, varying quantities of both black-and-white and color sets may be produced on the same day. The present demand for television sets is such that the company can sell as many sets as it can produce. In fact, some thought has been given to the possibility of either expanding the capacity of the present production facilities or adding a second production shift.

THE PROFIT OBJECTIVE

The objective of the television manufacturer is to produce that quantity of black-and-white and color television sets that will maximize total contribution to profit and overhead[1] subject to the restraints imposed by the capacity of the existing assembly plant. This profit objective may be expressed algebraically by letting Q_B denote the quantity of black-and-white sets produced in one day and letting Q_C denote the quantity of color sets produced in one day. (Q_B and Q_C are equivalent to X_1 and X_2 in Chapter 2.) Since the unit contributions to profit and overhead for Q_B and Q_C are \$15 and \$45 respectively, the profit function is given by Z:

$$Z = 15Q_B + 45Q_C.$$

We must now determine the values of Q_B and Q_C that will maximize Z and still satisfy the restrictions imposed by the present production facilities of the assembly plant.

THE PRODUCTION RESTRAINTS

The manufacture of television receiving sets involves the assembly of approximately 1000 components to a metal chassis base and the insertion of the chassis and the speaker into a cabinet. The process consists of four essential steps: (1) subassembly, or the fabrication of major components; (2) chassis assembly, or the wiring and assembly of components to a metal chassis; (3) set assembly, or the insertion of the chassis and speaker into a cabinet; (4) final inspection and testing.

[1] Throughout the discussion of this problem the term "profit" will be used as an abbreviation for the expression "contribution to profit and overhead."

Subassemblies and Components

Most of the major components of television sets are prepared on separate subassembly lines prior to movement to the main assembly line. However, speakers, cabinets, and transformers are purchased from outside suppliers. With the exception of color cathode-ray picture tubes, the manufacturer always carries an inventory consisting of a sufficient number of components to satisfy the daily production requirements for both black-and-white and color television sets. The plant has a daily production capacity of only 50 color cathode-ray picture tubes, and technical restraints prevent the purchase of color tubes from outside suppliers. This restraint on the production of color television sets may be expressed by the following algebraic relationship, if we again let Q_C denote the number of color television sets manufactured in one day.

$$Q_C \leq 50 \qquad (1)$$

This inequality indicates that the number of units of Q_C must be less than or equal to 50, where 50 is the daily supply of color picture tubes.

The significance of this relation is shown graphically in Figure 3-1. The area bounded by the equation ($Q_C = 50$), the Q_B-axis, and the Q_C-axis is the area of technical feasibility with regard to the restraint imposed by the limitation on the supply of color picture tubes. (The reason for restricting this area to positive values of Q_B and Q_C is that negative values are not meaningful in an economic sense.) Any

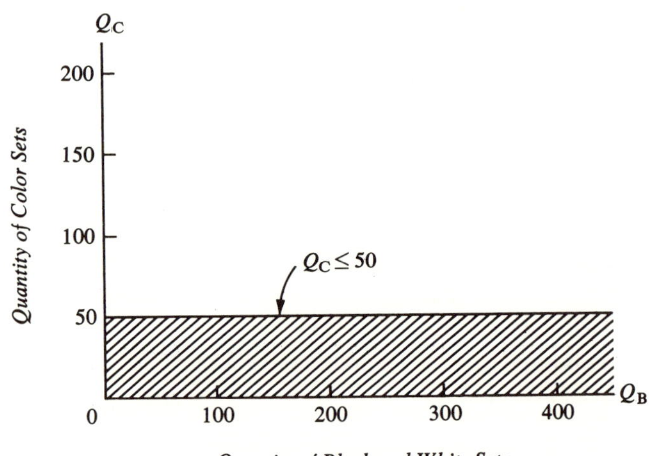

Figure 3-1. Production Restraints: Color Picture Tubes.

combination of values of Q_B and Q_C that lies either within or on the boundary of the shaded area of Figure 3-1 is a feasible solution to the problem in the sense that it satisfies the color picture-tube restraint. It will be noted that although Q_C may not exceed 50 units, there is no upper limit imposed on Q_B. However, we must now consider several additional restraints that will tend to reduce the area of technical feasibility and place an upper limit on Q_B.

Chassis Assembly

Each black-and-white set requires 6 man-hours of chassis assembly time, whereas a color set requires 18 man-hours. (A color set is considerably more complex than a black-and-white set.) The plant employs 225 men for an eight-hour shift to perform the chassis assembly operations. Since the plant is in operation for only one shift per day, the capacity of the chassis assembly line is (8 × 225) or 1800 man-hours per day.

The chassis assembly-line restraint may also be expressed mathematically by the use of our previous notation, i.e., by letting Q_B and Q_C denote respectively the number of black-and-white and color sets produced in a single day.

$$6Q_B + 18Q_C \leq 1800. \tag{2}$$

This relation indicates that the number of units of Q_B multiplied by the 6 man-hours that each unit of Q_B requires on the chassis assembly line plus the number of units of Q_C multiplied by the 18 man-hours that each unit of Q_C requires on the chassis assembly line must be less than or equal to 1800, where 1800 is the amount of labor capacity, measured in man-hours per day, available for chassis assembly operations.

The effect of this second production restraint is illustrated graphically in Figure 3-2. The area of technical feasibility has now been considerably reduced and takes the form of a convex polygon with vertices O, A, B, and C.[2]

[2] A convex polygon is a polygon of the form represented by (A) as opposed to that represented by (B). A vertex of a convex polygon is simply the intersection of two sides of a convex polygon. A convex polygon (or set) has the property that a straight line segment connecting any two points in the polygon (or set) lies entirely within the polygon (or set).

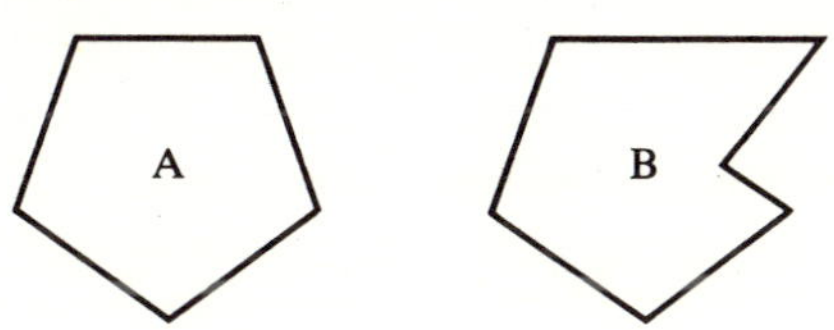

Convex *Nonconvex*

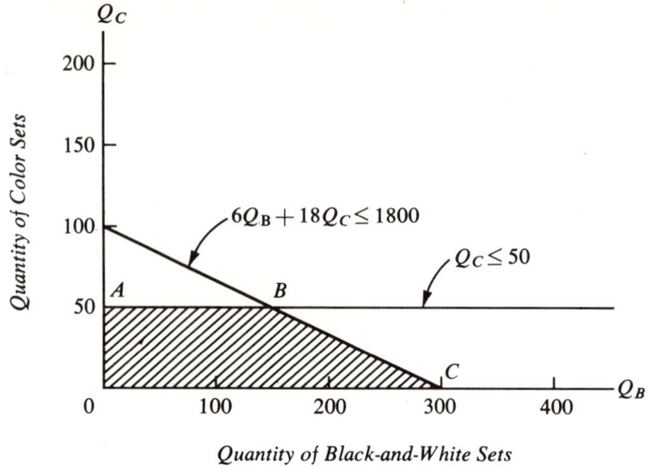

Figure 3-2. Production Restraints: (1) Color Picture Tubes and (2) Chassis Assembly.

Set Assembly

A black-and-white set must also spend 1 man-hour on the set assembly line, whereas a color set must spend 1.6 man-hours on this line. As there are 30 men working on the set assembly line, the daily capacity of this line is (8 × 30) or 240 man-hours of production. Algebraically, the set assembly-line restraint is given by

$$1.0Q_B + 1.6Q_C \leq 240. \tag{3}$$

The interpretation of this inequality is similar to that of restraint (2).

Figure 3-3 shows that the area of the convex polygon of technical feasibility has been further reduced by the introduction of the set assembly-line restraint. The set of feasible solutions to the problem has now been limited to those values of Q_B and Q_C that lie within or on the boundary of the convex polygon *OABCD*.

Final Inspection

Black-and-white sets each require 0.5 man-hours of testing and final inspection, whereas color sets require 2.0 man-hours. The plant employs 20 full-time inspectors and one part-time inspector. The part-time inspector spends only 2 hours per day inspecting television sets. Hence, the inspection capacity of the plant is (8 × 20.25) or 162 man-hours. Mathematically, the inspection restraint may be expressed as

$$0.5Q_B + 2.0Q_C \leq 162. \tag{4}$$

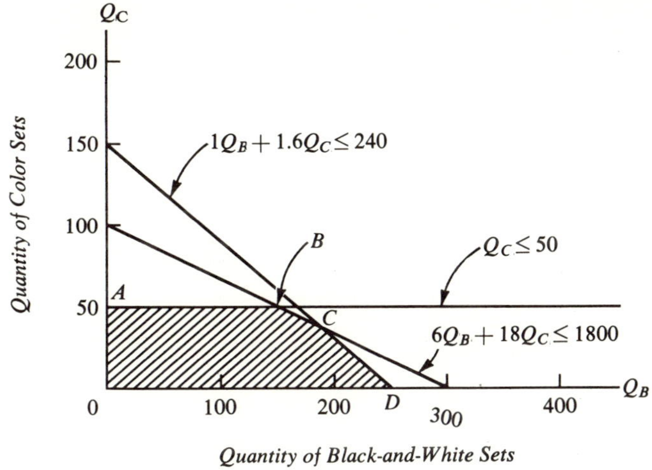

Figure 3-3. Production Restraints: (1) Color Picture Tubes; (2) Chassis Assembly; (3) Set Assembly.

THE FINAL SET OF FEASIBLE SOLUTIONS

Figure 3-4 shows the effect of all four production-capacity restraints on the problem. It is interesting to note that the inclusion of the fourth restraint causes the chassis-assembly restraint ($6Q_B + 18Q_C \leq 1800$) to become extraneous. The chassis-assembly restraint is no longer a boundary of the convex polygon of feasible solutions. This indicates that the convex polygon *OABCD* in Figure 3-4 is more restrictive on the solution than the chassis-assembly restraint.

The net effect of the production capacity restraints is to limit the number of alternative solutions available to the decision maker. However, the television manufacturer still has a relatively large number of alternatives from which to choose. Any combination within the area of the convex polygon of technical feasibility is a possible solution to the problem. The following combinations of Q_B and Q_C are technically feasible solutions; however, they do not by any means exhaust the list of feasible alternatives.

Q_B	Q_C
0	0
0	50
100	24
110	0
60	45

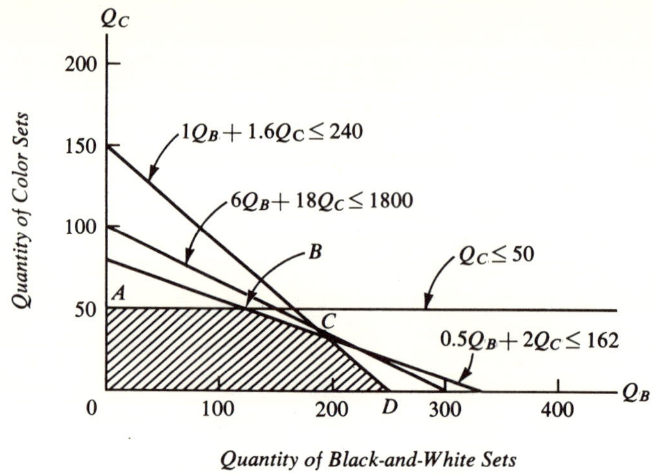

Figure 3-4. Production Restraints: (1) Color Picture Tubes; (2) Chassis Assembly; (3) Set Assembly; (4) Final Inspection.

THE OPTIMUM SOLUTION

Up to this point our efforts have been devoted to defining those combinations of black-and-white and color television sets that could be produced within the limits imposed by the production capacity of the plant. Having accomplished this objective, we can now return to our over-all objective—that is determine the particular combination of black-and-white and color television sets that will maximize total profits.[3]

The profit function, previously defined as

$$Z = 15Q_B + 45Q_C,$$

is a single-parameter family of parallel straight lines such that the value of Z increases as the line becomes farther removed from the origin. By expressing the profit function in the slope intercept form described in Chapter 1,

$$Q_C = -\frac{1}{3}Q_B + \frac{Z}{45},$$

we see that the slope of each line in this family of lines is equal to $-1/3$ and that Z will increase when Q_B and Q_C increase. This concept is represented graphically in Figure 3-5. Each line represents the locus of a set of combinations of Q_B and Q_C that yields

[3] It is possible to have more than one optimum combination of black-and-white and color television sets.

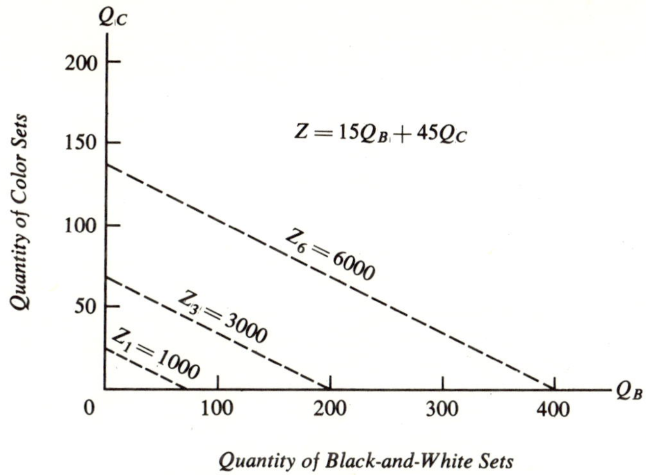

Figure 3-5. Iso-Profit Lines for Black-and-White and Color Sets.

equivalent profits. Hence, these lines are called iso-profit lines. For example, a profit of \$3000 may be obtained by producing any of the following combinations of Q_B and Q_C. (The number of combinations is certainly not limited to these four. Try a few other combinations.)

Combination	Q_B	*Profit at \$15 per unit*	Q_C	*Profit at \$45 per unit*	*Total profit*
1	200	\$3000	0	0	\$3000
2	140	\$2100	20	\$ 900	\$3000
3	50	\$ 750	50	\$2250	\$3000
4	20	\$ 300	60	\$2700	\$3000

The problem thus becomes one of determining that iso-profit line which is farthest away from the origin but still contains at least one point that is technically feasible. If we superimpose the iso-profit lines of Figure 3-5 on the convex polygon of technical feasibility shown in Figure 3-4, it can be seen in Figure 3-6 that none of these three iso-profit lines yields an optimum solution. The lines representing total profits of \$1000 and \$3000 are not as far from the origin as is technically possible, whereas the \$6000 profit line does not intersect the technically feasible area at any point and is therefore unobtainable. However, there is an additional line that is farthest away from the origin and still includes a point (point C) on the convex polygon $OABCD$. This is

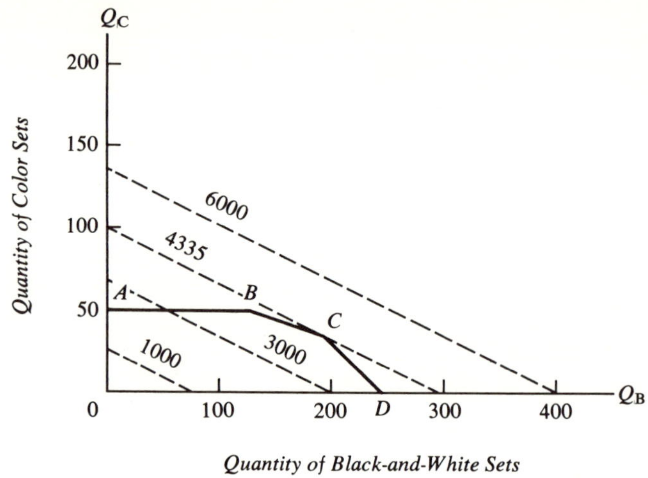

Figure 3-6. Graphical Solution Indicating Optimum Combination of Black-and-White and Color Sets.

the line of maximum profits. The line of maximum profits can be found in either of two ways:[4]

1. Plot the iso-profit line that passes through each vertex of the convex polygon. The line farthest away from the origin is the line of maximum profits. (The vertices can be found algebraically by solving the pair of simultaneous linear equations that intersect to form the particular vertex.)

2. For each vertex, calculate the profit for that particular combination of Q_B and Q_C. The vertex that yields the highest profit value lies on the line of maximum profits.

Either of these methods will yield a solution of point C on Figure 3-6. The following table provides conclusive evidence as to this fact.

[4] The optimum solution will normally be represented by one of the vertices of the convex polygon *OABCD*. However, in the special case where the slope of the iso-profit lines is equal to the slope of one of the restraints, there will not be a unique optimum solution. Although the graphical method illustrates that the solution to a linear programming problem will lie at a vertex, the only question is: which vertex? All methods of solving linear programming problems are ways of finding the optimum vertex without trying them all, as the number of possible vertices may be very large for most real-life problems.

Vertex	Q_B	*Profit at \$15 per unit*	Q_C	*Profit at \$45 per unit*	*Total profit*
O	0	0	0	0	0
A	0	0	50	\$2250	\$2250
B	124	\$1860	50	\$2250	\$4100
C	184	\$2760	35	\$1575	\$4335
D	240	\$3600	0	0	\$3600

The optimum solution therefore calls for the production of 184 black-and-white television sets and 35 color television sets. The profit resulting from this combination is \$4335. Although the solution to the television problem contains all integers (184 and 35), there is no reason to assume that the solution to a linear programming problem by the graphical or simplex method will necessarily yield results that are whole numbers. It is quite possible to have solutions that call for the production of, for example, 81.75 units of product. If a solution does not come out in whole numbers and the solution units are indivisible (as is the case with television sets), then one must devise some way of rounding the solutions to whole numbers with the inherent risk of obtaining a nonoptimum solution. The cost of this risk must be balanced against the cost of solving the problem by integer programming, a relatively inefficient computational procedure that guarantees integral solutions. Such decisions depend completely on the nature of the problem being considered, in this instance on the number of units involved and the value per unit.

Summary

The following is a brief summary of the graphical approach to linear programming:

1. State in equation form the objective function to be maximized.
2. List the linear restraints imposed on the objective function.
3. Represent the restraints graphically so as to form a polygon of technical feasibility.
4. Superimpose iso-profit lines on the polygon of technical feasibility and determine the optimum solution.

Although we have illustrated the graphical method by solving a problem involving profit maximization, the technique works equally well in solving minimization problems. In a cost minimization problem, one superimposes iso-cost lines on the polygon of technical feasibility and determines which iso-cost line is closest to the origin and still contains at least one point that is technically feasible.

Conclusions

The graphical method has only limited application to industrial problems owing to the inherent restraint placed on the number of variables that it is capable of handling. Although the graphical method can be extended to three-dimensional problems, in practice it is limited to problems involving only two real variables (for example, black-and-white and color television sets). The geometry of the three-dimensional case becomes so complex that considerable accuracy is likely to be sacrificed by using the graphical method to solve such problems. Problems involving more than three variables can obviously not be solved by purely graphical means.

Most industrial problems are so complex that they contain a multiplicity of variables. We must now investigate the more general technique for solving linear programming problems—the simplex method. Before taking up the simplex method, however, it is necessary to examine some properties and mechanics of simultaneous linear equations. This will be the subject of Chapter 4.

The graphical method should prove to be an invaluable pedagogical aid to understanding the simplex method, which appears in Chapter 5. Throughout the discussion of the simplex solution we shall refer to its relationship to the graphical solution. (The television assembly problem will also be solved by means of the simplex method.)

Problems

3-1. Use the graphical method to find the non-negative values of X_1 and X_2 that maximize $Z = 6X_1 + 8X_2$ subject to

$$2X_1 + 3X_2 \leq 12,$$

$$2X_1 + 1X_2 \leq 8.$$

3-2. Use the graphical method to find the non-negative values of X_1 and X_2 that minimize $Z = 10X_1 + 25X_2$ subject to

$$X_1 + X_2 \geq 50,$$
$$X_1 \geq 20,$$
$$X_2 \leq 40.$$

3-3. A manufacturer can produce two different products, A and B, during a given time period. Each of these products requires four different manufacturing operations: grinding, turning, assembly, and testing. The manufacturing requirements in hours per unit of product are given below for A and B.

	A	*B*
Grinding	1	2
Turning	3	1
Assembly	6	3
Testing	5	4

The available capacities of these operations in hours for the given time period are: grinding, 30; turning, 60; assembly, 200; testing, 200. The contribution to overhead and profit is \$2 for each unit of A and \$3 for each unit of B. The firm can sell all that it produces at the prevailing market price. Determine the optimum amount of A and B to produce during the given time period. (Use the graphical method.)

3-4. A scrap-metal dealer has received an order from a customer for at least 2000 pounds of scrap metal. The customer requires that at least 1000 pounds of the shipment of metal must be high quality copper that can be melted down and used to produce copper tubing. Furthermore, the customer will not accept delivery of the order if it contains more than 175 pounds of metal that he deems unfit for commercial use—that is, metal that contains an excessive amount of impurities and cannot be melted down and refined profitably.

The dealer can purchase scrap metal from two different suppliers in unlimited quantities with the following percentages (by weight) of high quality copper and unfit scrap.

	Supplier 1	*Supplier* 2
Copper	25%	75%
Unfit (*scrap*)	5%	10%

The costs per pound of metal purchased from Supplier #1 and Supplier #2 are 10¢ and 40¢ respectively. Determine the optimum quantities of metal for the dealer to purchase from each of the two suppliers.

3-5. The Johnson Manufacturing Company has two processes for producing widgets. It can produce a widget by combining 5 units of labor with 1 machine unit, or by combining 1 unit of labor with 2 units of machine capacity. Each widget contributes 1 unit toward profit and overhead, and the Johnson Company has 90 units of machine capacity and 200 units of labor available. Determine the production plan that maximizes profit.

3-6. Solve the Jones Company case at the end of the book.

4 Simultaneous Linear Equations

Since solving simultaneous equations is fundamental to the simplex method, we shall pause in this chapter to review some basic concepts of linear equations. By this means we plan to lay the algebraic foundation for the simplex method discussion in the next chapter.

First, we shall discuss the conditions under which linear equations have (1) a unique solution, (2) an infinite number of solutions, and (3) no solution. Then the Gauss-Jordan elimination method for solving linear equations is explained. The concept of a basic solution and its importance in linear programming is the next topic. The final sections deal with the derivation of a set of rules for generating basic solutions.

CONSISTENCY AND INDEPENDENCE

Consider the linear equation we graphed in Chapter 1:

$$4X_1 - 8X_2 = 10. \tag{1}$$

Now, we add a second equation and require that it be satisfied simultaneously with equation (1):

$$1X_1 + 0.5X_2 = 1. \tag{2}$$

Graphing the equations together in Figure 4-1, we obtain two intersecting straight lines.

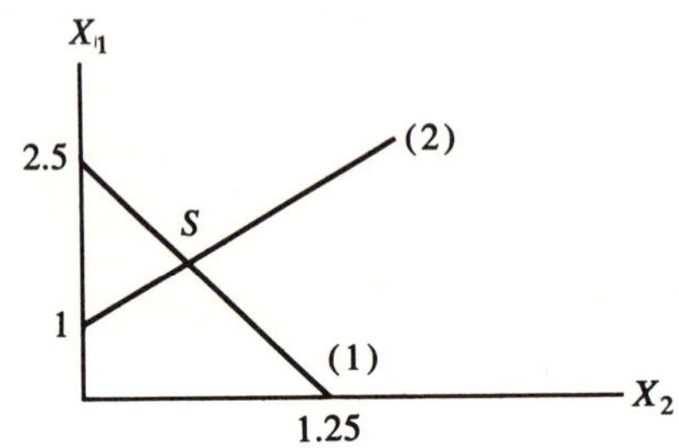

Figure 4-1. A Unique Solution.

By the simultaneous solution of the two equations we mean the pairs of values of X_1 and X_2 which satisfy both equations simultaneously. It is obvious from the graph that only *one* pair of values of X_1 and X_2 can satisfy both equations, that is, the values of X_1 and X_2 which correspond to the intersection point S. The solution is said to be *unique*.

Notice that we have *two* equations in *two* variables. Does it always hold true that two equations in two variables (or n equations in n variables) will yield a *unique* solution? The answer is *no*.

What if equation (2) had been

$$2X_1 + 4X_2 = 5? \tag{2'}$$

Graphing equations (1) and (2′), we obtain the results shown in Figure 4-2. The two equations coincide and appear as a single straight line. But what is the solution? Each

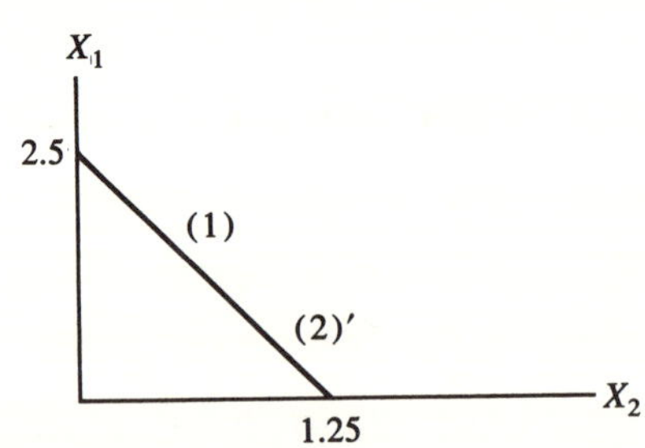

Figure 4-2. An Infinite Number of Solutions.

and every pair of values of X_1 and X_2 on the line is a solution. That is, there is an *infinite* number of solutions in this case, even though we still have two equations and two variables. What distinguishes this case from the first case, where the solution was *unique*? The answer is that equations (1) and (2′) are not *independent*. Thus, equation (2′) "adds no new information"; it can be obtained simply by multiplying both sides of equation (1) by 0.5.

One final case is possible. Let equation (2) now become

$$2X_1 + 4X_2 = 6. \tag{2''}$$

As depicted graphically in Figure 4-3, we get two parallel lines. In this third case there

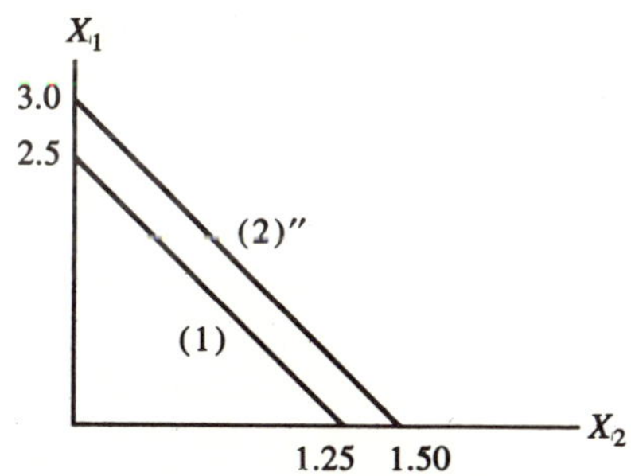

Figure 4-3. No Solution.

is *no* solution. No pair of values of X_1 and X_2 can possibly satisfy both equations. Equations (1) and (2″) are said to be *inconsistent*.

In summary, even given an equal number of variables and equations, a unique solution is not guaranteed. There will be an *infinite* number of solutions if the equations are *not independent*, and there can be *no* solution if the equations are *inconsistent*. Only if the equations are both independent and consistent are we assured a unique solution.

GAUSS-JORDAN ELIMINATION

We now consider a method for solving a system of linear equations. The Gauss-Jordan elimination method can be applied to very large systems and is well suited for computer computation. However, before illustrating the method, we need to recall two mathematical properties which will be used.

(A) Division of each term of an equation by a constant does not alter the solution to the equation system.

(B) Multiplication of each term of one equation by a constant, and then addition of the result to a second equation, does not alter the solution to the equation system.

Let us return to our earlier example.

$$4X_1 + 8X_2 = 10, \tag{1}$$

$$1X_1 - 0.5X_2 = 1. \tag{2}$$

The object is to use the two facts to transform equations (1) and (2) so that the solution is obvious. The solution will be obvious if the coefficients of the variables can be transformed into an *identity matrix*. An identity matrix is simply a square array of the numbers 0 and 1 in the form[1]

$$\begin{bmatrix} 1 & 0 \\ 0 & 1 \end{bmatrix}.$$

The first row of the matrix should be viewed as the coefficients of X_1 and X_2 in equation (1), and the second row should be similarly associated with equation (2). The solution will be obvious, once the original system is transformed into the identity matrix form. The value of X_1 will be the constant on the right-hand side of the first equation; and X_2 will equal the constant of the second equation.

Since the coefficient of X_1 in the first equation must be 1, we apply fact (A) to equation (1) by dividing through by 4. The result is

$$1X_1 + 2X_2 = 2.5. \tag{1a}$$

The coefficient of X_1 in equation (2) must become 0. If we apply fact (B) by multiplying equation (1a) by -1 and adding the result to equation (2), we get

$$0X_1 - 2.5X_2 = -1.5. \tag{2a}$$

The coefficients of equations (1a) and (2a) (*the coefficient matrix*) are

$$\begin{bmatrix} 1 & 2 \\ 0 & -2.5 \end{bmatrix}.$$

To complete the transformation to an identity matrix, we divide equation (2a) through by -2.5, obtaining

$$0X_1 + 1X_2 = 0.6. \tag{2b}$$

[1] This is a *second-order* identity matrix having two rows and two columns. We will need a *third-order* identity matrix later in this chapter.

Next, multiply (2b) by -2 and add the result to (1a). The resulting equation is

$$1X_1 + 0X_2 = 1.3. \tag{1b}$$

The coefficients of (1b) and (2b) now form the desired identity matrix:

$$1X_1 + 0X_2 = 1.3, \tag{1b}$$

$$0X_1 + 1X_2 = 0.6. \tag{2b}$$

The *unique* solution is $X_1 = 1.3$, $X_2 = 0.6$. This, of course, corresponds to point S in Figure 4-1.

We can now make a general statement about the solution of a system of n linear equations in n variables. Such a system will have a *unique* solution *if and only if* the system can be transformed into the identity matrix form.

It is interesting to observe what happens in the Gauss-Jordan procedure when (1) an infinite number of solutions, or (2) no solution occurs.

The equations of our previous example of an infinite number of solutions are

$$4X_1 + 8X_2 = 10 \tag{1}$$

and

$$2X_1 + 4X_2 = 5. \tag{2'}$$

Dividing (1) through by 4 gives

$$1X_1 + 2X_2 = 2.5. \tag{1a}$$

Multiplying the above equation by -2 and adding the result to (2′) produces

$$0X_1 + 0X_2 = 0. \tag{2a'}$$

Clearly, we cannot continue toward the identity matrix, since there is no way to get 1 as the coefficient of X_2 in the second equation. Equation (2a′) has the interesting property that *any* value of X_2 will satisfy it. Choose any arbitrary value for X_2, substitute that value in (1a), and a value of X_1 is thereby determined. Since X_2 was chosen arbitrarily, we can choose an infinite number of values, and, for each X_2, equation (1a) will imply a value for X_1. This outcome was shown graphically in Figure 4-2.

The equations in our earlier example where no solution is possible are

$$4X_1 + 8X_2 = 10 \tag{1}$$

and

$$2X_1 + 4X_2 = 6. \tag{2''}$$

Proceeding as before with the Gauss-Jordan method, we obtain

$$1X_1 + 2X_2 = 2.5, \qquad \text{(1a)}$$

$$0X_1 + 0X_2 = 1. \qquad \text{(2a'')}$$

Again, we are stopped from reaching the identity matrix form. However, in this instance we have obtained an equation, (2a″), which is clearly inconsistent. It is impossible to select values of X_1 and X_2 which will satisfy the equation. Hence, no solution to the system exists, as depicted graphically in Figure 4-3.

Although we have been working with the simple case of two equations in two variables, the results hold identically for any other system where the number of equations equals the number of variables.

TYPICAL LINEAR PROGRAMMING EQUATION SYSTEM

The system of equations in linear programming is always characterized by the number of variables exceeding the number of equations. This is sometimes called a system of *under-determined* equations. Although we have not explicitly discussed systems where the number of equations is less than the number of variables, we have done so in effect. The case discussed earlier, where an infinite number of solutions is possible, is actually an under-determined system. To understand why this is so, recall that the second equation in our example was simply a multiple of the first. It, therefore, "added nothing new" and could have been dropped from the system. If dropped, the first equation should be viewed as a system of one equation in two variables, or an under-determined system.

Consider the equations

$$4X_1 + 8X_2 + 2X_3 = 10 \qquad \text{(3)}$$

and

$$2X_1 + X_2 + 3X_3 = 2. \qquad \text{(4)}$$

There are an infinite number of solutions to the equations above. To see why, arbitrarily set X_3 equal to 1. The equations then become

$$4X_1 + 8X_2 = 10 - 2(1) = 8,$$
$$2X_1 + X_2 = 2 - 3(1) = -1.$$

We are now back to two equations in two variables. And, since the equations are independent and consistent, a unique solution can be found. But our choice of

$X_3 = 1$ was arbitrary. We could set X_3 equal to *any* number and then solve the two equations in two variables for X_1 and X_2. Hence, an infinite number of solutions is possible. Of course, we could just as easily have set either X_1 or X_2 equal to some value and solved for the other two variables.

BASIC SOLUTIONS

An important subset of the set of all possible solutions is the *set of basic solutions.*[2] A basic solution for our example, equations (3) and (4), is obtained by setting one variable equal to *zero* and solving for the other two. Thus, three basic solutions to equations (3) and (4) could be obtained by setting X_1, X_2, and X_3, in turn, equal to zero and solving for the other two variables. The basic solutions are given below:

1. $X_1 = 0$ — $X_2 = 1.18$, $X_3 = 0.27$,
2. $X_2 = 0$ — $X_1 = 3.25$, $X_3 = -1.5$,
3. $X_3 = 0$ — $X_1 = 0.5$, $X_2 = 1$.

To generalize, let m be the number of equations and n be the number of variables. The typical linear programming problem is characterized by $n > m$. A basic solution is obtained by setting $n - m$ variables equal to zero and solving the resulting system of m equations in m variables for the values of the m variables. Of course, we shall find a basic solution only if the resulting m equations in m variables are independent and consistent.

The maximum number of possible basic solutions can be determined by reference to the law of combinations. The number of ways of choosing $n - m$ variables from n variables, in order to set them equal to zero, is given by

$$\frac{n!}{(n-m)!\,m!},$$

where $n!$ is read "n factorial" and means $n(n-1)(n-2)\cdots 1$. For example, in the two-equation, three-variable illustration, where $n = 3$ and $m = 2$, there are three

[2] A set is simply any well-defined collection of distinct objects. The players on a particular baseball team, for example, would be a set, and the set of pitchers on the team would be a subset of the set.

possible basic solutions:

$$\frac{3!}{(3-2)!\,2!} = \frac{3 \cdot 2 \cdot 1}{(1) \cdot (2 \cdot 1)} = 3.$$

And, as shown above, all three do in fact exist.

IMPORTANCE OF BASIC SOLUTIONS IN LINEAR PROGRAMMING

To serve as an example, we shall select a three-equation, five-variable problem. One reason for using the larger system is to provide a slightly more complex illustration of the Gauss-Jordan method. Another reason is to examine the mathematics of the television assembly linear programming problem described in the last chapter. The discussion should also provide insight into the simplex method, which will be used to solve the same problem in the next chapter.

Recalling the three inequality restraints in Chapter 3, we had

$$Q_C \leq 50,$$

$$1.0Q_B + 1.6Q_C \leq 240,$$

$$0.5Q_B + 2.0Q_C \leq 162.$$

It was also pointed out that Q_C and Q_B must be non-negative, for it is not meaningful to speak of producing negative quantities of television sets.

The set of feasible solutions to the problem was shown by the shaded region in Figure 3-4 in Chapter 3. It is possible to convert the three inequalities into a system of three equations *which will have the identical set of solutions.* One reason for the conversion is simply that the algebraic method for solving equations is easier than the one for solving inequalities. The trick is to add *slack variables* to the inequalities. For example, adding a slack variable, S_1, to the color-tube restraint gives

$$Q_C + S_1 = 50.$$

In order to preserve the inequality, S_1 must be restricted to non-negative values. For example, $S_1 = -5$ would imply $Q_C = 55$, which is impossible. S_1 can be interpreted as the number of color tubes *not* used in production. Further discussion of slack variables will be included in the next chapter.

Adding slack variables to the three inequalities yields the following system of equations:

$$0Q_B + \quad 1Q_C + 1S_1 + 0S_2 + 0S_3 = \quad 50, \tag{5}$$

$$1Q_B + 1.6Q_C + 0S_1 + 1S_2 + 0S_3 = 240, \tag{6}$$

$$0.5Q_B + 2.0Q_C + 0S_1 + 0S_2 + 1S_3 = 162. \tag{7}$$

We must also add the restriction that the five variables be non-negative:

$$Q_B, Q_C, S_1, S_2, S_3 \geq 0.$$

As a result, we have three equations in five variables, or an under-determined system of equations. The set of feasible (non-negative) solutions is infinite and can be shown graphically as the shaded region in Figure 4-4. Since Q_C and Q_B are on the

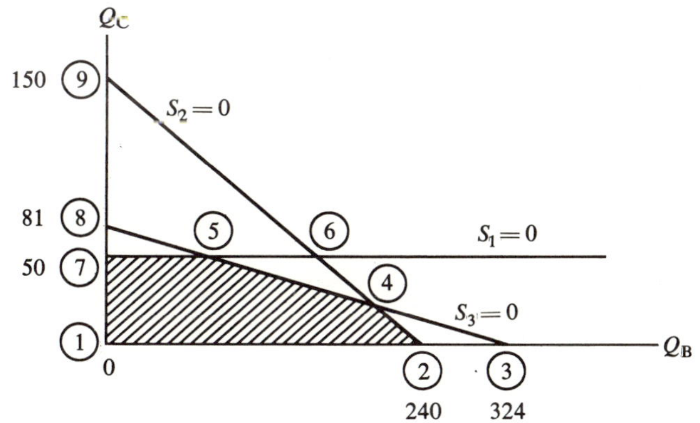

Figure 4-4. Set of Feasible Solutions and Basic Solutions.

axes, it is easy to understand the geometrical interpretation of their values. However, what is the interpretation of S_1, S_2, and S_3 in Figure 4-4? The three equations were graphed by setting their respective slack variables equal to zero. Hence, any point *on* the horizontal line, the graph of equation (5), implies that $S_1 = 0$. Similarly, points on the other lines imply that S_2 and S_3 equal zero. Each line in the figure is identified by the slack variable associated with its equation.

Which of the infinite solutions in the figure are basic solutions? Although the formula given earlier indicates ten *possible* solutions, there are only nine basic solutions, and they are numbered ①–⑨. Notice that each basic solution corresponds to an intersection point. If one studies the figure carefully, it is apparent that each intersection point has the property that two of the five variables are equal to zero. But, of course, this is the definition of a basic solution. That is, in the five-variable, three-equation case, set two variables equal to zero and solve the three-variable, three-equation system for the values of the other three variables.

Observe, also, that not all nine basic solutions are feasible. In fact, there are only five *basic feasible solutions*. These are numbers ①, ②, ④, ⑤, and ⑦. *Graphically, the basic feasible solutions correspond to the vertices of the feasible regions.* And, as we pointed out in Chapter 3, the optimum is always at a vertex. We can now say that the optimum solution will be a basic feasible solution.

This last statement is of profound importance and is normally referred to as the *fundamental theorem of linear programming*. It is valuable for solving linear programming problems because it is not necessary to examine each of the infinite feasible solutions but only the basic feasible solutions.

The simplex method is a search procedure that compares the value of the objective function for various basic feasible solutions. Clearly, solving a linear programming problem would be much more difficult if *all* feasible solutions had to be examined. In fact, as is discussed in the next chapter, the logic of the simplex method makes it unnecessary to examine *all* basic feasible solutions, and this reduces the computational problem even more.

We now turn to the algebraic process for determining the basic feasible solutions. First, we find the basic solutions. The formula for the maximum number of basic solutions for our case (where $m = 3$, $n = 5$) gives

$$\frac{5 \cdot 4 \cdot 3 \cdot 2 \cdot 1}{(2 \cdot 1)(3 \cdot 2 \cdot 1)} = 10.$$

In order to be systematic, we have set forth the ten possibilities in Table 4-1. The numbers correspond to the solutions as numbered in Figure 4-4. Notice that there is no solution (10) in the figure. The reason will be explained later.

The table shows the ten combinations of setting two variables out of the five equal to zero. To fill in the blanks requires in each case a simultaneous solution of three equations in three variables. To find the first basic solution, Q_C and Q_B are both set equal to zero in equations (5)–(7). Since Q_C and Q_B are equal to zero, we can eliminate them from the equations. This yields

$$1S_1 + 0S_2 + 0S_3 = 50,$$

$$0S_1 + 1S_2 + 0S_3 = 240,$$

$$0S_1 + 0S_2 + 1S_3 = 162.$$

TABLE 4-1

Solution Number	Q_C	Q_B	S_1	S_2	S_3
1	0	0			
2	0			0	
3	0				0
4				0	0
5			0		0
6			0	0	
7		0	0		
8		0			0
9		0		0	
10	0		0		

The Gauss-Jordan method of solution requires a transformation of the coefficient matrix above into the identity matrix. Since it is already in that form, we merely read the solution as $S_1 = 50$, $S_2 = 240$, and $S_3 = 162$. These values, together with $Q_B = Q_C = 0$, comprise the first basic solution.

To find solution 4, for example, we set $S_2 = S_3 = 0$, in equations (5) through (7) and obtain three equations in three variables. Although we could drop S_2 and S_3 from the equations, we will not do so. It is necessary to carry them along in the simplex method, and it seems appropriate to begin the practice here.

$$0\,Q_B + 1\,Q_C + 1S_1 + 0S_2 + 0S_3 = 50, \tag{8}$$

$$1\,Q_B + 1.6Q_C + 0S_1 + 1S_2 + 0S_3 = 240, \tag{9}$$

$$0.5Q_B + 2.0Q_C + 0S_1 + 0S_2 + 1S_3 = 162. \tag{10}$$

The coefficients of Q_B, Q_C, and S_1 are *not* in identity matrix form. First, we must interchange equations (8) and (9), since there is no way in the present arrangement to get the necessary 1 for the coefficient of Q_B.

After interchanging (8) and (9) and adding -0.5 times (9) to (10), we have

$$1Q_B + 1.6Q_C + 0S_1 + 1\,S_2 + 0S_3 = 240, \qquad (9)$$

$$0Q_B + 1\,Q_C + 1S_1 + 0\,S_2 + 0S_3 = 50, \qquad (8)$$

$$0Q_B + 1.2Q_C + 0S_1 - 0.5S_2 + 1S_3 = 42. \qquad (10)$$

To get zeros as the Q_C coefficients in (9) and (10) add -1.6 times (8) to (9) and add -1.2 times (8) to (10).

$$1Q_B + 0Q_C - 1.6S_1 + 1\ S_2 + 0S_3 = 160, \quad (9)$$

$$0Q_B + 1Q_C + 1\ S_1 + 0\ S_2 + 0S_3 = 50, \quad (8)$$

$$0Q_B + 0Q_C - 1.2S_1 - 0.5S_2 + 1S_3 = -18. \quad (10)$$

Then, divide (10) by -1.2 to get

$$0Q_B + 0Q_C + 1S_1 + \frac{5}{12}S_2 - \frac{5}{6}S_3 = 15. \quad (11)$$

Next, add 1.6 times (11) to (9) and -1 times (11) to (8) to get zeros as coefficients of S_1 in equation (9) and (8), respectively. In identity matrix form we now have

$$1Q_B + 0Q_C + 0S_1 + \frac{20}{12}S_2 - \frac{16}{12}S_3 = 184, \quad (9)$$

$$0Q_B + 1Q_C + 0S_1 - \frac{5}{12}S_2 + \frac{10}{12}S_3 = 35, \quad (8)$$

$$0Q_B + 0Q_C + 1S_1 + \frac{5}{12}S_2 - \frac{10}{12}S_3 = 15. \quad (11)$$

The basic solution is now easily seen to be

$$Q_B = 184,$$

$$Q_C = 35,$$

$$S_1 = 15,$$

$$S_2 = 0,$$

$$S_3 = 0.$$

Thus far we have encountered no difficulty in filling in the blanks of Table 4-1, since each of the resulting systems of three equations in three variables has yielded a unique solution. In fact, we can obtain nine basic solutions in this way. You are urged to apply the Gauss-Jordan method to the case where $Q_C = S_1 = 0$ (solution number 10 in Table 4-1). The three-equation, three-variable system turns out to be inconsistent; therefore no solution is possible. Although it is *possible* to obtain ten basic solutions when $n = 5$ and $m = 3$, only nine exist in this particular problem. Table 4-2 gives the complete list of basic solutions and indicates those that are feasible (that is, non-negative). You should relate these algebraic results to their graphical counterparts shown in Figure 4-4.

TABLE 4-2. Basic Solutions.

Solution Number	Q_C	Q_B	S_1	S_2	S_3	
1	0	0	50	240	162	Feasible
2	0	240	50	0	42	Feasible
3	0	324	50	−84	0	
4	35	184	15	0	0	Feasible
5	50	124	0	36	0	Feasible
6	50	160	0	0	−18	
7	50	0	0	160	62	Feasible
8	81	0	−31	110.4	0	
9	150	0	−100	0	−138	

To review our results thus far, we shall rewrite the original equations in tabular form. Rather than placing the variables after the coefficients, we shall simply place the appropriate variable as a column heading and omit the equalities. The right-hand side constants are shown in the column headed by P_0.

Solution 1 $(Q_B = Q_C = 0)$

Solution	Q_B	Q_C	S_1	S_2	S_3	P_0
S_1	0	1	1	0	0	50
S_2	1	1.6	0	1	0	240
S_3	0.5	2	0	0	1	162

We began by interpreting the equation above as a basic solution where $Q_B = Q_C = 0$. Recall that the coefficients of the other variable are in the proper identity matrix form for us to read off the solution. Each row (or equation) is given the variable name according to the solution it specifies. Thus, the first row indicates the solution of S_1 to be 50, the second row shows $S_2 = 240$, and the third row yields $S_3 = 162$.

The next result described was the basic solution where $S_2 = S_3 = 0$. We used the Gauss-Jordan method to transform the coefficients of Q_B, Q_C, and S_1 into identity matrix form, yielding the solution below:

Solution 4 ($S_2 = S_3 = 0$)

Solution	Q_B	Q_C	S_1	S_2	S_3	P_0
Q_B	1	0	0	$^{20}/_{12}$	$-^{16}/_{12}$	184
Q_C	0	1	0	$-^{5}/_{12}$	$^{10}/_{12}$	35
S_1	0	0	1	($^{5}/_{12}$)	$-^{10}/_{12}$	15

GENERATING NEW BASIC SOLUTIONS

Now we shall illustrate the algebraic process which is the essence of the simplex method. Given, say, basic solution 4, how can we generate a *new* basic solution wherein *one* of the zero variables in solution 4 is permitted to become nonzero? We want to make *either* S_2 or S_3 nonzero and, simultaneously, make either Q_B or Q_C or S_1 zero. Of course, we must maintain *two* zero variables in order for the new solution to be basic. That is why we must make one of the nonzero variables zero when we make a zero variable nonzero. Henceforth, we shall refer to a zero variable as a *nonbasic variable* and a nonzero variable as a *basic variable*. (Sometimes we shall refer to the basic variables as the "solution variables," since they appear in the solution column.)

In the next chapter we shall discuss how one chooses the *particular* nonbasic variable to enter the solution and how one chooses the *particular* basic variable to leave.[3] For the present, assume that we have selected S_2 *to enter* the solution and S_1 *to leave*. That is, we want to find the basic solution in which $S_1 = S_3 = 0$. The graphical representation is that we are moving from Solution (or vertex) 4 to Solution 5 in Figure 4-4.

The new solution must have an identity matrix associated with its basic variables Q_B, Q_C, and S_2. Referring back to Solution 4, we need to transform the equations so as to get a 1 for the coefficient of S_2 in the third equation (where the coefficient is now $^5/_{12}$). We have circled this coefficient and shall refer to it as the *pivot element*. Notice that the pivot element lies in the entering variable column and the leaving variable row. We must also obtain zeros for the coefficients of S_2 in the first and second rows.

[3] There are two major considerations involved in this decision. One is to select as the entering variable that variable which will increase the value of the objective function at the greatest rate. The other consideration is to make the selection such that the resulting basic solution will also be *feasible*.

The first step in deriving Solution 5 is to get a 1 in the place of the pivot element. The Gauss-Jordan method calls for division of that row by the pivot element, or by $5/12$. The result is shown below, where we have also replaced S_1 by S_2 in the solution column.

Solution	Q_B	Q_C	S_1	S_2	S_3	P_0
Q_B	1	0	0	$20/12$	$-16/12$	184
Q_C	0	1	0	$-5/12$	$10/12$	35
S_2	0	0	2.4	1	-2	36

Bear in mind that the object is to obtain an identity matrix as the coefficient matrix of the new basic variables, Q_B, Q_C, and S_2. To complete this transformation we must get zeros in column S_2, rows Q_B and Q_C.[4]

Multiply the S_2 row in the table above by $-20/12$ and add the result to the Q_B row. Then, multiply the S_2 row by $5/12$ and add the result to the Q_C row. You should be familiar with these Gauss-Jordan procedures by now. The result is given below:

Solution 5 ($S_1 = S_3 = 0$)

Solution	Q_B	Q_C	S_1	S_2	S_3	P_0
Q_B	1	0	-4	0	②	124
Q_C	0	1	1	0	0	50
S_2	0	0	2.4	1	-2	36

The solution is $Q_B = 124$, $Q_C = 50$, $S_2 = 36$, $S_1 = 0$, and $S_3 = 0$.

Finally, to summarize how one basic solution can be generated from another, we list the following rules.

RULES FOR GENERATING NEW BASIC SOLUTIONS

Let E be the nonbasic variable which is to enter the solution (become basic) and let L be the basic variable which is to leave (become nonbasic).

1. The pivot element lies in the L row and the E column of the old solution table.

[4] It should be understood that there is no need for the three variables associated with the identity matrix to be adjacent. Although the identity matrix would perhaps "look better" if we exchanged columns S_1 and S_2, there is no mathematical reason for doing so.

2. Divide each number in the L row of the old table by the pivot element. The result is the E row in the new table.
3. For all other rows, say row K, take the negative of the number in row K of the old table and in the pivot element column. Multiply this times the E row of the *new* table and add the result to the K row to obtain the K row for the new table.

To illustrate the rules, we take one last example. We want to derive Solution 7 from Solution 5. That is, we select S_3 as the entering variable and Q_B as the leaving variable.

Rule 1. The pivot element is the 2 in row Q_B and column S_3 of Solution 5.

Rule 2. Division of row Q_B by 2 yields the new row S_3:

Solution	Q_B	Q_C	S_1	S_2	S_3	P_0
S_3	½	0	−2	0	1	62

Rule 3. Multiply row S_3 by 0 and 2, respectively, and add the results to rows Q_C and S_2 respectively to obtain new rows Q_C and S_2:

Row Q_C	Q_B	Q_C	S_1	S_2	S_3	P_0
0 *times* S_3 *row*	0	0	0	0	0	0
+ *old* Q_C *row*	0	1	1	0	0	50
New Q_C *row*	0	1	1	0	0	50

Row S_2	Q_B	Q_C	S_1	S_2	S_3	P_0
2 *times* S_3 *row*	1	0	−4	0	2	124
+ *old* S_2 *row*	0	0	2.4	1	−2	36
New S_2 *row*	1	0	−1.6	1	0	160

Place the new rows in the Solution-7 table:

Solution 7 ($S_1 = Q_B = 0$)

Solution	Q_B	Q_C	S_1	S_2	S_3	P_0
S_3	½	0	−2	0	1	62
Q_C	0	1	1	0	0	50
S_2	1	0	−1.6	1	0	160

CONCLUSIONS

The most important conclusions reached in this chapter are listed below:

1. The inequality restraints of linear programming problems can be written as equations which have the same feasible solutions as the inequalities.

2. Linear programming restraint equations are characterized by $n > m$, where n is the number of variables and m is the number of equations. Such equations have an infinite number of solutions.

3. Basic solutions to the equations are determined by setting $n - m$ variables equal to zero and solving the resulting m equations in m variables for the values of the m variables. There is a finite number of basic solutions.

4. Not all basic solutions are feasible; those basic solutions with all variables non-negative are basic feasible solutions.

5. Basic feasible solutions are the algebraic analog of the vertices in the feasible region of the graphical analysis.

6. The fundamental theorem of linear programming states that the optimum solution will be a basic feasible solution.

7. The Gauss-Jordan method can be used to derive a set of rules for calculating a new basic solution from an old basic solution. The new

basic solution differs from the old in that a nonbasic variable in the old becomes a basic variable in the new, and vice versa.

In the next chapter we shall make use of the rules for generating new basic solutions in an exposition of the simplex method. We have not yet discussed how the particular entering and leaving variables should be chosen. The answer to this question is the essence of the simplex method.

5 The Simplex Method

The simplex method is an algebraic procedure that, through an iterative process, progressively approaches and ultimately reaches an optimum solution. This technique is applicable to any problem that can be formulated in terms of a linear objective function subject to a set of linear restraints. Although there is no theoretical restriction placed on the number of variables or restraints contained in a simplex problem, practical considerations may place an upper limit on the number of variables and restraints.

The simplex method may be summarized briefly as follows:

1. Define the problem in terms of a linear objective function subject to a set of linear restraints.
2. Introduce slack and artificial variables to convert inequalities into equalities and provide a basic feasible solution to the problem.
3. Construct the initial simplex table for a basic feasible solution.
4. Determine which variable should be entered into the solution (become basic), if the solution is not optimum.
5. Determine which variable should leave the solution (become non-basic).

6. Construct a new simplex table reflecting the changes that have been made in the solution.
7. Repeat steps 4 through 6 until an optimum solution is reached, i.e., until the solution cannot be further improved.

To illustrate the simplex method of solving linear programming problems, we shall consider two specific industrial problems. The first of these problems is the television assembly problem, which was solved by the graphical method in Chapter 3. The second problem involves the blending of raw materials for the production of soap.

THE TELEVISION ASSEMBLY PROBLEM

Define the Problem

In Chapter 3 we defined the problem of the television manufacturer as one of producing that quantity of black-and-white and color television sets which would maximize total contribution to profit and overhead subject to the restraints imposed by the capacity of the existing assembly plant. Algebraically the problem can be stated as one of determining that quantity of Q_B and Q_C which will maximize the profit function

$$Z = 15Q_B + 45Q_C,$$

subject to the following restraints imposed by the present production facilities:

color picture tube supply, $Q_C \leq 50;$

set assembly capacity, $1.0Q_B + 1.6Q_C \leq 240;$

final inspection capacity, $0.5Q_B + 2.0Q_C \leq 162.$

(Note that the chassis assembly restraint is not included, as it was previously shown to be extraneous. Although it could have been included, it would complicate the example needlessly. However, this does not in any way imply that the simplex method is dependent upon the graphical method for the elimination of extra restraints.)

Introduce Slack Variables

Since the simplex method requires that the linear restraints be expressed as equalities rather than as inequalities, slack variables must be introduced into the expressions for production capacity restrictions. Since there are three inequalities, three slack variables (S_1, S_2, and S_3) must be introduced.

$$1Q_C + 1S_1 = 50, \tag{1}$$

$$1.0Q_B + 1.6Q_C + 1S_2 = 240, \tag{2}$$

$$0.5Q_B + 2.0Q_C + 1S_3 = 162. \tag{3}$$

Technically, S_1 corresponds to the quantity of color picture tubes that are not used in the production of television sets. Computationally, S_1 represents a fictitious product that requires one color picture tube per unit of product. Equation (1) expresses the fact that the number of color picture tubes used in the production of color sets plus the number of color picture tubes not used is equal to the number of color picture tubes available.

Slack variables S_2 and S_3 correspond to idle man-hours of set assembly and final inspection capacity, respectively. They might be said to represent fictitious products moving through the set assembly line and the final inspection line, respectively. Equation (2) states that the number of man-hours devoted to the production of black-and-white sets on the set assembly line, plus the number of man-hours devoted to the production of color sets on the set assembly line, plus the slack variable or idle set assembly-line capacity is equal to the total number of man-hours available on the set assembly line. Equation (3) states that the number of man-hours devoted to the final inspection of black-and-white sets, plus the number of man-hours devoted to the final inspection of color sets, plus the slack variable or idle final inspection capacity is equal to the total number of man-hours of final inspection capacity available.

In order that all variables are represented in each equation, we add real and slack variables with zero coefficients to those equations that are lacking in some variables. For example, $0Q_B$, $0S_2$, and $0S_3$ are added to the first equation.

$$0Q_B + 1.0Q_C + 1S_1 + 0S_2 + 0S_3 = 50, \tag{1}$$

$$1.0Q_B + 1.6Q_C + 0S_1 + 1S_2 + 0S_3 = 240, \tag{2}$$

$$0.5Q_B + 2.0Q_C + 0S_1 + 0S_2 + 1S_3 = 162. \tag{3}$$

With the introduction of the three slack variables S_1, S_2, and S_3, the objective function must be altered accordingly. The variables S_1, S_2, and S_3, having a unit contribution to profit and overhead of zero, must be added to the objective function. (No real production implies no contribution to profit and overhead.)

$$Z = 15Q_B + 45Q_C + 0S_1 + 0S_2 + 0S_3.$$

Our problem may now be stated as one of solving a three-equation, five-variable system of linear equations. As we have learned in Chapter 4, such a system has an infinite number of solutions. But, according to the fundamental theorem of linear programming, we should expect the optimum solution to be a basic feasible solution (or a vertex of the feasible region).[1] The number of basic feasible solutions is finite and we shall use the objective function to indicate which of these possible solutions is preferable.

The simplex method may start with any basic feasible solution. The procedure is then to generate new basic feasible solutions which increase the value of the objective function. This iterative process continues until, ultimately, the maximum (optimum) basic feasible solution is selected.

In the television assembly problem it is convenient to begin with a particular basic feasible solution, namely the solution in which we produce nothing at all. The advantage of this starting point is that it is both technically feasible and relatively easy to formulate. If we produce nothing, then Q_B and Q_C are both equal to zero, and we have 50 unused color picture tubes, 240 man-hours of idle set assembly-line capacity, and 162 man-hours of idle final inspection capacity. This corresponds to the solution at the origin in Figure 4-4, or to Solution 1 in Table 4-2. Hence, the initial basic feasible solution is

$$\begin{aligned} Q_B &= 0, \qquad & S_1 &= 50, \\ Q_C &= 0, \qquad & S_2 &= 240, \\ & & S_3 &= 162. \end{aligned}$$

Construct the Initial Simplex Table

It is more convenient for computational purposes to express the relationships between profits, variables, and restraints in the form of a table rather than to retain the expressions in equation form. The initial simplex table for the basic feasible solution discussed above is represented by Table 5-1.

Note that the elements of rows 1, 2, and 3 of Table 5-1 correspond to the coefficients of the variables in the three production restraint equations. For example,

[1] In the graphical method it was stated that in the special case where the slope of the iso-profit lines is equal to the slope of one of the restraints, there will be a range of optimum solutions, rather than a unique solution. Even so, a vertex or basic feasible solution will always be equivalent to any of the boundary points.

TABLE 5-1

Row	C_i	Column / C_j / Solution	1 / 15 / Q_B	2 / 45 / Q_C	3 / 0 / S_1	4 / 0 / S_2	5 / 0 / S_3	P_0
1	0	S_1	0	(1.0)	1	0	0	50
2	0	S_2	1.0	1.6	0	1	0	240
3	0	S_3	0.5	2.0	0	0	1	162
		Z_j	0	0	0	0	0	0
		$C_j - Z_j$	15	45	0	0	0	

the elements in row 1 correspond to the coefficients of the variables in Equation (1):

$$0Q_B + 1Q_C + 1S_1 + 0S_2 + 0S_3 = 50.$$

The elements in the "solution" column indicate the three basic variables (S_1, S_2, and S_3) that are contained in the initial basic solution. (Recall from Chapter 4 that these are the variables which are associated with the identity matrix.) The values of the three solution variables are found in the P_0 column. (We shall use the terms "basic variables" and "solution variables" interchangeably.)

The production requirements for producing one black-and-white set and one color set are found in Columns 1 and 2, respectively. Column 1 indicates that the production of one black-and-white set requires 1.0 man-hours of set assembly-line capacity and 0.5 man-hours of final testing capacity. Column 2 shows that the production of one color set requires 1.0 color picture tubes, 1.6 man-hours of set assembly-line capacity, and 2.0 man-hours of testing capacity. Although Columns 3, 4, and 5 may be interpreted in a manner similar to that of Columns 1 and 2, very little is gained by such an interpretation. For example, Column 3 indicates that the production of 1 unit of color picture-tube excess capacity requires 1.0 color picture tubes (an obvious tautology).

The elements in the C_j row (where the j subscript refers to the column number) are the coefficients of the variables in the objective function, that is, the gross contributions to profit and overhead per unit for each variable Q_B, Q_C, S_1, S_2, and S_3. The elements in the C_i column (where the i subscript refers to the row number) are the contributions to profit and overhead per unit for each of the three basic variables

S_1, S_2, and S_3, in the initial basic feasible solution. The fact that the C_i's are all equal to zero in the initial solution indicates that it is not profitable to have the plant lie completely idle. That is, total profit is

$$Z = 0(50) + 0(240) + 0(162) = 0.$$

Under each column is a Z_j total. Notice that total profit, as calculated above, is Z_0 (the value of Z_j in Column P_0 of Table 5-1). The Z_j total of the other columns is the amount of gross profit that is forfeited by introducing one unit of the variable in the j column into the solution.

For example, suppose that we want to produce one color set Q_C. The production requirements for one color set are given in Column 2:

1.0 color picture tube;

1.6 man-hours of assembly capacity;

2.0 man-hours of testing capacity.

Thus, the basic variables in the solution (S_1, S_2, and S_3) would each be reduced by the amounts above. The value of S_1 would be reduced from 50 to 49, S_2 would be reduced from 240 to 238.4, and S_3 would be reduced from 162 to 160. Altogether, the value of profit would be reduced by

$$C_1(1) + C_2(1.6) + C_3(2.0),$$

where C is the profit function coefficient of S_1, S_2, and S_3 (the C_i). In this example, the C_i's are all equal to zero, so that

$$Z_2 = 0(1.0) + 0(1.6) + 0(2.0) = 0.$$

We find that the production of one unit of Q_C has the opportunity cost of using resources that were previously used in the production of S_1, S_2, and S_3. The opportunity cost in this case is zero, because reducing production of S_1, S_2, and S_3 has no effect on profit. However, this is not the complete story. The color set directly contributes \$45 to profit (as shown in the C_j row), so that the *net effect* on profit is $C_2 - Z_2$, or \$45 − 0 = \$45. Thus, the $C_j - Z_j$ row shows the net effect on profit of producing one additional unit of the jth variable.

If we let a_{ij} denote the element in the ith row and jth column of the table, the Z_j's (including Z_0) are defined by

$$Z_j = \sum_{i=1}^{3} C_i a_{ij} \qquad (j = 0, 1, 2, 3, 4, 5);$$

$$\begin{aligned}
Z_1 &= 0(0) + 0(1.0) + 0(0.5) = 0,\\
Z_2 &= 0(1.0) + 0(1.6) + 0(2.0) = 0,\\
Z_3 &= 0(1) + 0(0) + 0(0) = 0,\\
Z_4 &= 0(0) + 0(1) + 0(0) = 0,\\
Z_5 &= 0(0) + 0(0) + 0(1) = 0,\\
Z_0 &= 0(50) + 0(240) + 0(162) = 0.
\end{aligned}$$

And, the $C_j - Z_j$ row of Table 5-1 is

$$\begin{aligned}
C_1 - Z_1 &= 15 - 0 = 15,\\
C_2 - Z_2 &= 45 - 0 = 45,\\
C_3 - Z_3 &= 0 - 0 = 0,\\
C_4 - Z_4 &= 0 - 0 = 0,\\
C_5 - Z_5 &= 0 - 0 = 0,\\
C_0 - Z_0 &= 0 - 0 = 0.
\end{aligned}$$

Determine Which Variable to Introduce into the Solution

Having completely defined the initial simplex table, we must now determine whether an optimum solution has been achieved. Stated differently, we want to find out whether or not the value of the profit function will be increased by introducing some quantity of Q_B or Q_C (the two nonbasic variables) into the solution. The answer to this question may be found in the $C_j - Z_j$ row, for it indicates that there would be a gain of \$15 for each black-and-white set produced and \$45 for each color set produced. It further indicates that nothing would be gained at this point by increasing the excess capacity of the plant, that is, by increasing the quantities of S_1, S_2, and S_3 that are presently in the solution. As long as there is at least one $C_j - Z_j$ that is positive (greater than zero), an optimum solution has not been reached. The solution can be improved by introducing into the solution some quantity of that variable which has the highest positive value in the $C_j - Z_j$ row: the variable with the greatest net unit profit. Although profit can be increased by introducing some positive quantity of any variable with a positive value in the $C_j - Z_j$ row, one criterion for selecting which variable to introduce into the solution is to select the variable with the highest unit contribution to profit and overhead. However, there are numerous other criteria for making this decision, some of which yield an optimum solution in fewer steps but require more computation between steps. An example of such an approach is the decision rule that introduces the variable contributing most to *total* profit rather than the variable with the highest *unit* contribution to profit. The reader may wish to try both approaches and compare the results.

Since the simplex technique requires that variables be entered into the solution one at a time, a decision must be reached as to which variable to introduce first. Again we refer to the $C_j - Z_j$ row and find that \$45 represents the largest net unit profit in the row. Hence, we shall introduce as many units of Q_C into the solution as is technically feasible, for total profits will be increased by \$45 for each color set produced. In Step 5 we shall determine how much Q_C can be introduced into the solution, or, how many color sets can be produced if only color sets are produced.

Determine Which Variable to Remove from the Solution

In the initial solution the variable Q_C was assumed to be equal to zero, and thus was not one of the three basic variables contained in the solution. However, in Step 4 we decided to enter some positive quantity (as much as possible) of Q_C into the solution. Since there are three restraint equations, there must be exactly three variables with non-negative values in the solution. Hence, one variable must leave the solution to permit the entrance of Q_C, that is one variable must be set equal to zero (become nonbasic).

This is equivalent to saying that we shall introduce the maximum amount of Q_C that is technically feasible. But this feasibility is determined by the number of units of excess capacity, that is the values of S_1, S_2, and S_3. In other words, units of Q_C will be produced up to and including the point where either (1) all of the available color picture tubes have been used ($S_1 = 0$), (2) there is no longer any set assembly-line capacity available ($S_2 = 0$), or (3) there is no final testing capacity available ($S_3 = 0$). When the maximum amount of Q_C has been entered into the solution, either S_1, S_2, or S_3 will be equal to zero, depending on which restraint is most limiting. Thus, we shall still have three equations with exactly three solution variables.

For each unit of Q_C brought into the solution, 1.0 units of S_1 (color picture tubes), 1.6 units of S_2 (set assembly-line capacity), and 2.0 units of S_3 (testing capacity) must be removed (subtracted) from the solution. These numbers are called coefficients of exchange, for in effect, we are exchanging 1.0 units of S_1, 1.6 units of S_2, and 2.0 units of S_3 for one unit of Q_C.

In this problem no more than 50 units of S_1, nor 240 units of S_2, nor 162 units of S_3 may be removed from the solution. This is equivalent to saying that no negative production is possible. Dividing these capacity limits by the appropriate exchange coefficients will yield the following three limits to the number of units of Q_C that can be brought into the solution:

$$\frac{50}{1.0} = 50,$$

$$\frac{240}{1.6} = 150,$$

$$\frac{162}{2.0} = 81.$$

It can be seen that 50 is the smallest capacity restraint and, therefore, the largest number of units of Q_C that can be brought into the solution. The production of 50 color television sets removes all of the S_1 (color picture tubes) from the solution, for there were only 50 color picture tubes available in the first place. In the graphical solution this is equivalent to going out along the Q_C-axis from the origin until the first capacity restraint is reached. This corresponds to Solution 7 in Figure 4-4 and Table 4-2.

Construct a New Simplex Table

A new simplex table must now be constructed that will reflect the changes caused by introducing 50 units of Q_C into the solution and removing 50 units of S_1. This is equivalent to saying that we must now generate a new basic solution where Q_C is the entering variable and S_1 is the leaving variable. (We have already ensured that the new basic solution will be *feasible* by the selection of S_1 as the leaving variable.)

The new solution will induce changes in the exchange coefficients, the P_0 column, the C_i column, and the Z_j and $C_j - Z_j$ rows. The values of these new coefficients must be computed in order to test for an optimum solution. If the new solution is not optimum, the procedure must be repeated until an optimum solution is found.

Since the variable Q_C is being introduced into the solution and variable S_1 is being removed from the solution, Q_C will be substituted for S_1 in the solution column of the simplex table (Table 5-2a). The new variable contributes \$45 per unit to profit, and hence the C_i value for the first row becomes \$45.

TABLE 5-2a

		Column	1	2	3	4	5	
		C_j	15	45	0	0	0	
Row	C_i	*Solution*	Q_B	Q_C	S_1	S_2	S_3	P_0
1	45	Q_C	0	1.0	1	0	0	50
2	0	S_2						
3	0	S_3						

In order to calculate the exchange coefficients for rows 1, 2, and 3, we simply apply the rules for generating new basic solutions which were derived at the end of the last chapter.

Rule 1. The pivot element is the circled 1.0 in row 1, column 2 of Table 5-1. (It is located in the entering variable column and the leaving variable row.)

Rule 2. Division of row 1 of Table 5-1 by 1.0 yields the new row 1 shown in Table 5-2a.

Rule 3. Multiply row 1 in Table 5-2a by -1.6 and -2.0 (negatives of numbers in the pivot element column, rows 2 and 3 of Table 5-1) and add the results to rows 2 and 3 respectively of Table 5-1. This gives the two new rows 2 and 3 which appear in Table 5-2b:

Row 2	Q_B	Q_C	S_1	S_2	S_3	P_0
-1.6 times row 1 (Table 5-2a)	0	-1.6	-1.6	0	0	-80
+ old row 2 (Table 5-1)	1.0	1.6	0	1	0	240
New row 2 (Table 5-2b)	1.0	0	-1.6	1	0	160

Row 3	Q_B	Q_C	S_1	S_2	S_3	P_0
−2.0 times row 1 (Table 5-2a)	0	−2.0	−2.0	0	0	−100
+ old row 3 (Table 5-1)	0.5	2.0	0	0	1	162
New row 3 (Table 5-2b)	0.5	0	−2.0	0	1	62

We can now enter the new coefficients in Table 5-2b.

TABLE 5-2b

		Column	1	2	3	4	5	
		C_j	15	45	0	0	0	
Row	C_i	*Solution*	Q_B	Q_C	S_1	S_2	S_3	P_0
1	45	Q_C	0	1	1.0	0	0	50
2	0	S_2	1.0	0	−1.6	1	0	160
3	0	S_3	(0.5)	0	−2.0	0	1	62
		Z_j	0	45	45	0	0	2250
		$C_j - Z_j$	15	0	−45	0	0	

One additional comment on the coefficients in Table 5-2b is necessary. Two negative exchange coefficients appear in column 3. These may cause some confusion if the significance of a negative coefficient is not understood. As we have already seen, a positive exchange coefficient in a simplex table indicates that quantity of a variable which must be removed from the solution in order to introduce a unit of some variable not presently in the solution. Referring to column 1 in Table 5-2b, we can see that in order to introduce 1 unit of Q_B (that is, produce one black-and-white television set) it is necessary to remove 0 units of Q_C—since, obviously, color picture tubes are not used in the production of black-and-white sets—1 unit of S_2, and ½ unit of S_3. This, too, is obvious, since each unit of Q_B requires 1 man-hour of set-assembly capacity and ½ man-hour of final inspection capacity.

On the basis of the logic presented above, we are now led to the conclusion that a negative exchange coefficient requires not that something be removed from the solution, but rather, that something be brought into the solution. Let us see why this is so. Column 3 of Table 5-2b shows a negative 1.6 and a negative 2.0. Thus, if the $C_j - Z_j$ row indicated that some quantity of S_1 should be introduced into the solution, meaning that some slack color picture tube capacity would be created, what exactly would be involved? Clearly for each unit of S_1 brought into the solution one color set is removed, as indicated by the positive exchange coefficient (1.0 in row 1, column 3). Since one color set is now removed from the solution, 1.6 man-hours of S_2 (set assembly capacity), and 2.0 man-hours of S_3 (final inspection capacity) are automatically created, that is, brought into the solution. We therefore conclude that although the magnitude of a positive exchange coefficient is a limiting factor on the number of units of a new variable that can be introduced into the solution, the existence of a negative exchange coefficient is not a limiting factor since some product quantities presently in the solution are to be increased rather than decreased.

The C_i's in Table 5-2b are 45, 0, and 0 for rows 1, 2, and 3, respectively. The Z_j row includes

$$
\begin{aligned}
Z_1 &= 45(0) + 0(1.0) + 0(5) = 0,\\
Z_2 &= 45(1) + 0(0) + 0(0) = 45,\\
Z_3 &= 45(1) + 0(-1.6) + 0(-2) = 45,\\
Z_4 &= 45(0) + 0(1) + 0(0) = 0,\\
Z_5 &= 45(0) + 0(0) + 0(1) = 0,\\
Z_0 &= 45(50) + 0(160) + 0(62) = 2250.
\end{aligned}
$$

The value of Z_0 is actually the value of the profit function when

$$
\begin{aligned}
Q_C &= 50, \qquad Q_B = 0,\\
S_2 &= 160, \qquad S_1 = 0;\\
S_3 &= 62;\\
Z &= 15Q_B + 45Q_C + 0S_1 + 0S_2 + 0S_3\\
&= 15(0) + 45(50) + 0(0) + 0(160) + 0(62),\\
Z &= \$2250.
\end{aligned}
$$

The $C_j - Z_j$ row is computed in the same manner as before.

Repeat Steps 4 through 6 Until an Optimum Solution is Obtained

Upon completion of the second simplex table we must again ask whether an optimum solution has been obtained. A quick glance at the $C_j - Z_j$ row of Table 5-2b shows that each unit of Q_B introduced into the solution at this point will increase profits by \$15. Since there is a positive quantity in the $C_j - Z_j$ row, an optimum solution has not been found.

In order to determine the maximum amount of Q_B that can be introduced into the solution, it is necessary to divide the exchange coefficients given for Q_B (black-and-white sets) into the production quantities in the P_0 column. This calculation appears below.

$$\frac{50}{0} = \infty,$$

$$\frac{160}{1.0} = 160,$$

$$\frac{62}{0.5} = 124.$$

From these figures it is evident that it is physically impossible to introduce more than 124 units of Q_B into the solution, and produce more than 124 black-and-white sets. This is due to the fact that there are only 62 man-hours of S_3 (final testing capacity) available, and each black-and-white set requires 0.5 man-hours of final testing.

Hence, a third simplex table must be formulated reflecting the effects of introducing 124 units of Q_B into the solution and removing 62 units of S_3. The pivot element in Table 5-2b now becomes the circled 0.5. Applying Rules 1–3 for generating new basic solutions, we obtain the coefficients in rows 1, 2, and 3 of Table 5-3.

The computational procedures for developing the elements in the Z_j and $C_j - Z_j$ rows are identical to those of the previous two simplex tables.

In Table 5-3 the values of the solution variables are

$$Q_C = 50,$$

$$S_2 = 36,$$

$$Q_B = 124.$$

TABLE 5-3

		Column	1	2	3	4	5	
		C_j	15	45	0	0	0	
Row	C_i	Solution	Q_B	Q_C	S_1	S_2	S_3	P_0
1	45	Q_C	0	1	1	0	0	50
2	0	S_2	0	0	(2.4)	1	−2	36
3	15	Q_B	1	0	−4	0	2	124
		Z_j	15	45	−15	0	30	4110
		$C_j - Z_j$	0	0	15	0	−30	

This solution corresponds to Solution 5 in Figure 4-4 and Table 4-2. We began at the origin, moved up the Q_C axis to Solution 7, and have now moved horizontally to Solution 5. Each time, the value of the profit function increased. Thus, the current value of profit is \$4110.

$$Z = 15(124) + 45(50) = \$4110.$$

However, Table 5-3 also fails to yield an optimum solution, as it is possible even yet to improve the solution by introducing the maximum amount of S_1 that is technically feasible. As can be seen below, the maximum amount of S_1 that can be brought into the solution is 15 units because of the limited supply of S_2.

$$\frac{50}{1} = 50,$$

$$\frac{36}{2.4} = 15,$$

$$\frac{-4}{124} = -32.$$

All 36 man-hours of S_2 (set-assembly capacity) are thereby removed from the solution and replaced by 15 idle color picture tubes, S_1.

Once again, applying the rules for generating new basic solutions, a fourth simplex table can be derived where the pivot element in Table 5-3 is the circled 2.4.

Table 5-4 indicates that there are no longer any positive elements in the $C_j - Z_j$ row. Since all of these elements are either equal to zero or negative, the value of the objective function cannot be increased by introducing a new variable into the

TABLE 5-4

		Column	1	2	3	4	5	
		C_j	15	45	0	0	0	
Row	C_i	*Solution*	Q_B	Q_C	S_1	S_2	S_3	P_0
1	45	Q_C	0	1	0	−5/12	10/12	35
2	0	S_1	0	0	1	5/12	−10/12	15
3	15	Q_B	1	0	0	20/12	−16/12	184
		Z_j	15	45	0	6.25	17.50	(4335)
		$C_j - Z_j$	0	0	0	−6.25	−17.50	

solution. Hence, an optimum solution has been achieved. From Table 5-4 the optimum solution is found to be

$$Q_C = 35,$$

$$S_1 = 15,$$

$$Q_B = 184,$$

which corresponds to Solution 4 in Figure 4-4.

The value of the profit function is

$$Z = \$15(184) + \$45(35) + 0(15) + 0(0) + 0(0) = \$4335.$$

However, the value of the optimum solution should come as no surprise to us, since it is identical with the results achieved in Figure 3-6 of the graphical method. As in the graphical solution, the optimum solution provided by the simplex method calls for the production of 184 black-and-white sets and 35 color sets, yielding a total contribution to profit and overhead of \$4335.

To check the technical feasibility of this solution, we can simply substitute the values of the solution variables back into the original production restraint equations, as follows:

$$0Q_B + 1.0Q_C + 1S_1 + 0S_2 + 0S_3 = 50, \quad (1)$$
$$0(184) + 1.0(35) + 1(15) + 0(0) + 0(0) = 50,$$
$$50 = 50.$$

$$1.0Q_B + 1.6Q_C + 0S_1 + 1S_2 + 0S_3 = 240, \quad (2)$$
$$1.0(184) + 1.6(35) + 0(15) + 1(0) + 0(0) = 240,$$
$$240 = 240.$$

$$0.5Q_B + 2.0Q_C + 0S_1 + 0S_2 + 1S_3 = 162, \quad (3)$$
$$0.5(184) + 2.0(35) + 0(15) + 0(0) + 1(0) = 162,$$
$$162 = 162.$$

Since all three of the production restraint equations are satisfied, the solution is technically feasible.

THE VALUE OF A MARGINAL UNIT OF PRODUCTION CAPACITY

The variable S_2 has a $C_j - Z_j$ value, or net-contribution-to-profit value, of $-\$6.25$. This means that a unit of this variable (slack man-hours of set assembly-line capacity) introduced into the solution at this point would reduce the value of the profit function by \$6.25. Since the capacity of this line is fully utilized, introducing 1 hour of slack time corresponds to reducing the capacity of this line by 1 unit, which would reduce the optimum solution by \$6.25. If, in fact, the original set assembly-line capacity had been 1 man hour less than 240, the optimum solution would have a value of \$6.25 less. Likewise, if the original set assembly-line capacity were 1 man hour larger than the 240, the optimum solution would be increased by \$6.25. In other words, the net contribution to profit per unit at this stage represents the value of a marginal unit of set assembly-line capacity implied by the solution. These marginal profits are usually referred to as "shadow prices," or opportunity profits. For the managerial decision maker, opportunity profit is defined as the profit of foregone opportunities, or in other words a comparison between the alternative that was accepted and the alternative that was rejected. A decision rule in which opportunity profits are involved hinges on a comparison between what the decision maker or the firm is doing and what it could be doing; it is the difference between those alternatives that constitutes the relevant profit consideration. The value $-\$6.25$ is the difference between the alternative of *not* increasing set assembly-line capacity by 1 man hour and the alternative of increasing set assembly-line capacity by 1 man hour.

It is quite possible that the most important benefit to be derived from the solution of a linear programming problem is not the optimum solution itself, but rather the opportunity profits of the resources (color picture tubes, set assembly-line capacity, and final testing capacity) that are in limited supply and impose restraints on the variables of the problem. Although we are interested in finding an optimum solution to problems that have been formulated under a set of clearly defined conditions, we are even more interested in finding the value of altering these conditions. How would our profits be affected by an increase in the capacity of the set assembly line? Should we produce 50 color picture tubes each day? Should we make any changes in our final testing capacity? These are the questions of real economic significance that can be answered by examining the shadow prices of the final simplex table of an optimum solution to a linear programming problem. In an economic sense, the total value of the set assembly-line capacity is given by (\$6.25)(240) = \$1500. The opportunity profit of variable S_3 (slack final testing capacity) is \$17.50. The total value of the final testing capacity is (\$17.50)(162) = \$2835.

In total, these two values equal \$1500 + \$2835 = \$4335, which is identical to the value of the profit function for an optimum solution

$$\$45(35) + 0(15) + \$15(184) = \$4335.$$

In other words, with an optimum solution, the values imputed to the variables not in the final solution (S_2 and S_3) in total equal the value of the objective function. This leads us to the dual problem of linear programming, which we shall discuss in the next chapter.

In the following section we consider in detail a linear programming problem involving the minimization of an objective function.

THE SOAP BLENDING PROBLEM

A firm that manufactures soap products (hypothetical in this case) finds that it cannot satisfy the demand for one of its powdered laundry soap products during a particular week. The firm is confronted with a loss of sales, amounting to approximately 1000 pounds, simply because its production capacity is inadequate. Although this situation is only temporary, the firm does have a way whereby the excess demand can be satisfied.

Specifically, the firm is considering buying powdered soap from two other manufacturers and blending them into a soap that would at least meet the specifications of the soap presently being marketed by the company. That this is possible has been established through previous experience. There are two questions of economics associated with this possibility. First, what would be the optimum quantities of Soap number 1 and Soap number 2, the soaps produced by the two manufacturers, to purchase in order to satisfy demand and to provide a product that is identical with the present product? Second, would this be a profitable venture?

The company can purchase as much Soap number 1 and Soap number 2 as it needs for \$0.10 and \$0.08 per pound, respectively. It can sell the additional 1000 pounds of soap for \$200. However, it will incur additional costs of \$25 for blending, packaging, handling, and shipping the extra 1000 pounds of soap.

To the consumer the three most important characteristics of powdered laundry soap are washing power, sudsiness, and hardness. By analyzing certain chemical and physical characteristics of soap it is possible to actually measure the respective characteristics of a particular soap product. Extensive research has indicated that the particular soap product in question must have a washing power factor of at least 6.5, a sudsiness factor of at least 3.0, and a hardness factor no greater than 4.0.

The two soaps that are being considered for blending purposes have the following characteristics:

	Soap 1	*Soap* 2
Washing Power	6	9
Sudsiness	2	6
Hardness	1	5

The company's research department has determined that the characteristics of the two soaps being considered for blending combine linearly by weight when blended. In other words, if equal weights of Soap 1 and Soap 2 are combined, the resulting blend will have the following characteristics: washing power, 7.5; sudsiness, 4.0; and hardness, 3.0.

As can be seen, this particular blend has the required characteristics for the company's product. However, it may not be the ratio of Soap 1 to Soap 2 that yields the lowest possible cost. We may now turn our attention to this and other considerations.

Although the solution to this problem is trivial and may be obtained simply by inspection, it does serve to introduce several new concepts to the reader—namely: cost minimization, artificial variables, and the use of linear programming to solve blending problems. Furthermore, the type of reasoning used in formulating and solving this problem should prove useful in solving large-scale blending problems, such as the Midwest Grain Elevator case included at the end of the text.

Define the Linear Programming Problem

In order to determine the minimum cost blend of Soap 1 and Soap 2 that will satisfy demand (1000 pounds) and meet the product specifications required for this product, we may again resort to linear programming.

Objective Function. If we let X_1 denote the quantity of Soap 1 (in pounds) and X_2 denote the quantity of Soap 2 (in pounds) used in the blend, the objective of the firm is to minimize the cost function

$$Z = 0.10X_1 + 0.08X_2.$$

Restraints. The next part of the problem is to define the restraints imposed by the required product characteristics and the demand situation.

The washing power factor restraint may be expressed algebraically by the inequality

$$\frac{6X_1}{1000} + \frac{9X_2}{1000} \geq 6.5.$$

The expression $X_1/1000$ denotes that proportion of the blend having a washing power factor of 6.0, whereas $X_2/1000$ denotes that proportion of the blend having a washing power of 9.0. Thus, the inequality states that the proportions of Soap 1 and Soap 2 must be such that a blend will be produced whose washing power factor is at least 6.5.

The sudsiness and hardness restraints may be expressed in a similar manner:

$$\text{(sudsiness)} \qquad \frac{2X_1}{1000} + \frac{6X_2}{1000} \geq 3;$$

$$\text{(hardness)} \qquad \frac{1X_1}{1000} + \frac{5X_2}{1000} \leq 4.$$

A fourth restraint, which recognizes the demand situation, must be added. The total weight of the blend must be equal to the total weight demanded:

$$X_1 + X_2 = 1000.$$

When both sides of the washing power, sudsiness, and hardness restraints are multiplied by 1000, the four restraints become

$$6X_1 + 9X_2 \geq 6500,$$

$$2X_1 + 6X_2 \geq 3000,$$

$$1X_1 + 5X_2 \leq 4000,$$

$$X_1 + X_2 = 1000.$$

Introduce Slack and Artificial Variables

Again it becomes necessary to introduce slack variables to convert inequalities to equalities and to provide a basic feasible solution.

If we introduce a slack variable S_1, whose cost is zero, the washing power factor restraint becomes

$$6X_1 + 9X_2 - S_1 = 6500.$$

That is, the sum of the washing power quantities, $6X_1 + 9X_2$, may be equal to 6500 washing power-pounds if there is no slack, or it may exceed 6500 washing power-pounds requiring some positive value for S_1 to reduce the equation to 6500 washing power-pounds.[2]

However, for the basic feasible solution in the simplex method, the slack variable used for each equation must be non-negative. If we let X_1 and X_2 be equal to zero and then solve for S_1, we find that ($S_1 = -6500$). This negative solution violates our previous assumption concerning simplex solutions, that the simplex method only considers non-negative solutions. A second extra variable, known as an artificial variable, must therefore be inserted into the equation to provide a *non-negative*

[2] Washing power-pounds are the units that result from multiplying washing power by pounds.

basic feasible solution.[3] We will call the new variable A_1. If A_1, which provides a basic feasible solution, were to appear in the final solution it would mean that the blend would have a washing-power factor of less than 6.5. Since the manufacturer requires a washing-power factor of at least 6.5, we must ensure that A_1 does not appear in the final solution. This is accomplished by assigning to A_1, and any other artificial variable we may be using, a very high cost M, sufficient to drive A_1 out of the solution. (For example, we might assume $M = \$1{,}000{,}000$.) The zero cost associated with S_1 means that a washing-power factor greater than 6.5 in the blend is acceptable to the manufacturer. The washing-power restraint thus becomes

$$6X_1 + 9X_2 - S_1 + A_1 = 6500. \tag{1}$$

The sudsiness restraint is also a greater-than-or-equal-to ($\geq$) inequality and requires a slack variable, S_2, and an artificial variable, A_2. The slack variable, S_2, is assigned a zero cost and the artificial variable is assigned a very large cost, M. The sudsiness restraint then becomes an equality:

$$2X_1 + 6X_2 - S_2 + A_2 = 3000. \tag{2}$$

The hardness restraint is similar to the restraints in the television assembly problem in the sense that it is of the less-than-or-equal-to variety ($\leq$). A positive

[3] Another way to view the artificial variable is that it provides an identity matrix in the coefficient matrix for an initial basic solution. A two-restraint problem, such as,

$$1X_1 + 1X_2 \geq 4$$
$$2X_1 + 1X_2 \leq 3$$

may be converted into equations by adding slack variables, S_1 and S_2:

$$1X_1 + 1X_2 - 1S_1 + 0S_2 = 4$$
$$2X_1 + 1X_2 + 0S_1 + 1S_2 = 3$$

However, observe the coefficient matrix:

$$\begin{bmatrix} 1 & 1 & -1 & 0 \\ 2 & 1 & 0 & 1 \end{bmatrix}.$$

The coefficient of the slack variable, S_1, of the "greater than" inequality restraint has an unacceptable minus sign in that it fails to provide an identity matrix. However, adding A_1 to the first restraint enables us to obtain an identity matrix and, thereby, an initial basic solution. Thus,

$$\begin{bmatrix} 1 & 1 & -1 & 0 & 1 \\ 2 & 1 & 0 & 1 & 0 \end{bmatrix}.$$

slack variable, S_3, must be introduced at zero cost:

$$1X_1 + 5X_2 + S_3 = 4000. \qquad (3)$$

The demand restraint is somewhat different from the other restraints in that X_1 plus X_2 must be exactly equal to 1000 pounds. Neither positive nor negative slack will be tolerated. But if X_1 and X_2 are set equal to zero in order to obtain an initial basic feasible solution, we are confronted with the absurd result

$$0 + 0 = 1000.$$

In order to prevent the occurrence of such an absurdity and to provide an initial basic feasible solution to the problem, an artificial variable, A_3, must be introduced into the equation:

$$X_1 + X_2 + A_3 = 1000. \qquad (4)$$

Since no slack is permitted in the final solution, A_3 must also be assigned a very large positive cost, M. (The cost of failing to satisfy demand is assumed to be very high.)

With the addition of six new variables (S_1, A_1, S_2, A_2, S_3, A_3), the objective function must be adjusted accordingly. The objective function becomes

$$Z = 0.10X_1 + 0.08X_2 + 0S_1 + MA_1 + 0S_2 + MA_2 + 0S_3 + MA_3.$$

Construct the Initial Simplex Table

By setting X_1, X_2, S_1, and S_2 equal to zero and solving for A_1, A_2, S_3, and A_3, the initial basic feasible solution to this problem becomes

$$A_1 = 6500,$$
$$A_2 = 3000,$$
$$S_3 = 4000,$$
$$A_3 = 1000.$$

The initial simplex table (Table 5-7) may now be constructed according to the procedure described in the television assembly problem. With one minor exception, the necessary steps for reaching an optimum solution are identical to those in the television problem. Rather than selecting the variable with the largest positive $C_j - Z_j$ value to introduce into the solution next, we shall choose the variable that has the most negative $C_j - Z_j$ value, since the objective is to minimize total cost. The variable with the most negative $C_j - Z_j$ element will provide the largest net reduction

TABLE 5-5

	Column	1	2	3	4	5	6	7	8	
	C_j	0.10	0.08	0	M	0	M	0	M	
C_i	*Solution*	X_1	X_2	S_1	A_1	S_2	A_2	S_3	A_3	P_0
M	A_1	6	9	−1	1	0	0	0	0	6500
M	A_2	2	⑥	0	0	−1	1	0	0	3000
0	S_3	1	5	0	0	0	0	1	0	4000
M	A_3	1	1	0	0	0	0	0	1	1000
	Z_j	$9M$	$16M$	$-M$	M	$-M$	M	0	M	$10{,}500M$
	$C_j - Z_j$	$0.10–9M$	$0.08–16M$	M	0	M	0	0	0	

in cost per unit of variable introduced. An optimum solution will be reached when all of the elements in the $C_j - Z_j$ row are either positive or equal to zero, indicating that the total cost cannot be further reduced.[4]

The variable X_2 will be the first variable introduced into the solution, since $(0.08 - 16M)$, the $C_j - Z_j$ value of X_2, is more negative than $(0.10 - 9M)$, the $C_j - Z_j$ value of X_1. The variable to be introduced into the solution is indicated by the arrow at the bottom of each simplex table, whereas the variable to be removed from the solution is indicated by the arrow to the left of each table.

Determine the Optimum Solution

In order to reach an optimum solution, four iterations are required. These are indicated by Tables 5-8 through 5-11. In Table 5-11 it will be noted that all of the elements in the $C_j - Z_j$ row are either zero or positive. In other words, total cost cannot be further reduced by introducing any quantity of any variable not presently in the solution. Hence, an optimum solution has been achieved.

[4] This problem could also be solved by expressing costs as negative profits. The variable with the most positive $C_j - Z_j$ value would then be introduced into the solution first, as is the case with all maximization problems. When all of the elements in the $C_j - Z_j$ row are either negative or equal to zero, an optimum solution will have been reached. Many linear programming codes written for electronic computers require that costs be expressed as negative profits.

TABLE 5-6

		Column	1	2	3	4	5	6	7	8	
		C_j	0.10	0.08	0	M	0	M	0	M	
Row	C_i	Solution	X_1	X_2	S_1	A_1	S_2	A_2	S_3	A_3	P_0
1	M	A_1	3	0	-1	1	$\frac{3}{2}$	$-\frac{3}{2}$	0	0	200
2	0.08	X_2	$\frac{1}{3}$	1	0	0	$-\frac{1}{6}$	$\frac{1}{6}$	0	0	50
3	0	S_3	$-\frac{2}{3}$	0	0	0	$\frac{5}{6}$	$-\frac{5}{6}$	1	0	150
4	M	A_3	$\frac{2}{3}$	0	0	0	$\frac{1}{6}$	$-\frac{1}{6}$	0	1	50
		Z_j	$0.03 + \frac{11}{3}M$	0.08	$-M$	M	$-0.01 + \frac{10}{6}M$	$0.01 - \frac{10}{6}M$	0	M	40 + 2
		$C_j - Z_j$	$0.07 - \frac{11}{3}M$	0	$+M$	0	$0.01 - \frac{10}{6}M$	$-0.01 + \frac{16}{6}M$	0	0	

TABLE 5-7

		Column	1	2	3	4	5	6	7	8	
		C_j	0.10	0.08	0	M	0	M	0	M	
Row	C_i	Solution	X_1	X_2	S_1	A_1	S_2	A_2	S_3	A_3	P
1	0.10	X_1	1	0	$-\frac{1}{3}$	$\frac{1}{3}$	$\frac{1}{2}$	$-\frac{1}{2}$	0	0	200
2	0.08	X_2	0	1	$\frac{1}{9}$	$-\frac{1}{9}$	$-\frac{2}{6}$	$\frac{2}{6}$	0	0	250
3	0	S_3	0	0	$-\frac{2}{9}$	$\frac{2}{9}$	$\frac{7}{6}$	$-\frac{7}{6}$	1	0	17,5
4	M	A_3	0	0	$\frac{2}{9}$	$-\frac{2}{9}$	$-\frac{1}{6}$	$\frac{1}{6}$	0	1	50
		Z_j	0.10	0.08	$-0.02+\frac{2}{9}M$	$0.02-\frac{2}{9}M$	$0.02-\frac{1}{6}M$	$-0.02+\frac{1}{6}M$	0	M	89+
		C_j-Z_j	0	0	$0.02-\frac{2}{9}M$	$-0.02+\frac{11}{9}M$	$-0.02+\frac{1}{6}M$	$0.02+\frac{5}{6}M$	0	0	

TABLE 5-8

		Column	1	2	3	4	5	6	7	8	
		C_j	0.10	0.08	0	M	0	M	0	M	
Row	C_i	Solution	X_1	X_2	S_1	A_1	S_2	A_2	S_3	A_3	
1	0.10	X_1	1	0	0	0	$\frac{1}{4}$	$-\frac{1}{4}$	0	$\frac{3}{2}$	
2	0.08	X_2	0	1	0	0	$-\frac{1}{4}$	$\frac{1}{4}$	0	$-\frac{1}{2}$	
3	0	S_3	0	0	0	0	1	-1	1	1	2
4	0	S_1	0	0	1	-1	$-\frac{3}{4}$	$\frac{3}{4}$	0	$\frac{9}{2}$	
		Z_j	0.10	0.08	0	0	0.005	-0.005	0	0.11	
		$C_j - Z_j$	0	0	0	M	-0.005	$M + 0.005$	0	$M - 0.11$	

TABLE 5-9

	Column	1	2	3	4	5	6	7	8	
	C_j	0.10	0.08	0	M	0	M	0	M	
C_i	*Solution*	X_1	X_2	S_1	A_1	S_2	A_2	S_3	A_3	P_0
0.10	X_1	1	0	0	0	0	0	$-1/4$	$5/4$	250
0.08	X_2	0	1	0	0	0	0	$1/4$	$-1/4$	750
0	S_2	0	0	0	0	1	-1	1	1	2000
0	S_1	0	0	1	-1	0	0	$3/4$	$21/4$	1750
	Z_j	0.10	0.08	0	0	0	0	-0.005	0.10	85
	$C_j - Z_j$	0	0	0	M	0	M	0.005	$M - 0.10$	

It will also be observed that there are no artificial variables in the final solution. The variables A_1, A_2, S_3, and A_3, which were found in the initial solution, have been replaced by X_1, X_2, S_2, and S_1.

The final solution calls for a blend consisting of 250 pounds of Soap 1 and 750 pounds of Soap 2.[5] The total cost of the raw materials required to produce 1000 pounds of soap of the desired quality is \$85. The cost of blending, packaging, and handling the 1000 pounds of soap is \$25. Hence, the total cost of producing 1000 additional pounds of soap with the required attributes is \$25 plus \$85 or \$110. Since the company can sell this soap for \$200, it would appear that the purchase of 250 pounds of Soap 1 and 750 pounds of Soap 2 would be a favorable alternative. Naturally, the profitability of this alternative would have to be compared with the profitability of other alternatives before a final decision could be made. Other alternatives might include expanding the capacity of the soap plant, buying soap from other suppliers, or turning the account over to a competitor.

THE DEGENERATE CASE

When using the simplex method to solve linear programming problems, a rather peculiar solution known as a "degenerate" solution may arise, if one of the

[5] Again it should be noted that the simplex method provides no assurance of integral solutions. This problem was intentionally formulated in such a way as to minimize computational difficulty and thus provide an integral solution. Real-life problems do not have this characteristic. The cases at the end of the text will serve to illustrate this point.

restraints is redundant. For example, if there are two restraints, $X_1 \leq 15$ and $X_1 \leq 25$, the second restraint is not necessary. The restraint stating that X_1 must be less than or equal to 15 eliminates the need for the restraint that X_1 must be less than or equal to 25. Unfortunately, most redundant restraints are not as easily recognized as the one above.

Only certain redundant restraints will lead to degenerate solutions, and it is not always a simple matter to predict a degenerate situation. Degeneracy becomes apparent in the simplex calculation when more than one variable in the solution has the same lower figure limiting the number of units of the new variable to be brought into the solution. Since those solution variables that are "limiting" will all become equal to zero, if they are not already equal to zero, the question arises as to which variable to remove from the solution.

An acceptable rule to follow is to choose one of the limiting variables *arbitrarily*. The only problem with this practice is that it is theoretically possible to encounter *cycling*. That is, the simplex tables would begin to repeat themselves, and an optimum would not be reached. However, cycling is such a remote possibility that it is not necessary to follow the complex rules which have been derived to avoid cycling.[6]

A PROBLEM WITHOUT AN OPTIMUM SOLUTION

Under certain conditions it is possible for a linear programming problem not to have an optimum solution. Assume that the variable X_1 is to be introduced into the solution in a particular simplex problem because it either has the most positive value in the $C_j - Z_j$ row, if the problem involves profit maximization, or it has the most negative value in the $C_j - Z_j$ row, if the problem involves cost minimization. If all of the exchange coefficients in the X_1 column are negative and/or zero, the problem does not have an optimum solution. The negative coefficients indicate that for every unit of X_1 introduced into the solution the ability to introduce additional units will be increased. The more X_1 that is used the higher the profits (or the lower the costs). The problem does not have a unique optimum solution, since there is no limit to the amount of X_1 that can be introduced into the solution. Such a solution is called an *unbounded* solution.

[6] G. Hadley, *Linear Programming* (Reading, Mass.: Addison-Wesley Publishing Co., 1962) p. 174.

As a general rule, a linear programming problem with no unique optimum solution implies that one of the restraints has been incorrectly defined. If properly defined, most industrial problems will not result in a nonsolvable simplex table.

ALTERNATE OPTIMUM SOLUTIONS

If $C_j - Z_j \neq 0$ for all of the variables not included in the optimum solution of a linear programming problem (nonbasic variables), then a unique solution exists. However, if $C_j - Z_j = 0$ for some nonbasic variables, then alternate optimum solutions exist. These alternate optimum solutions can be determined by introducing into the solution those nonbasic variables whose $C_j - Z_j = 0$.

ADVANCED LINEAR PROGRAMMING

Although we have used only the algebra of linear equations in this chapter, the reader who wishes to become a serious student of linear programming must acquire a knowledge of vectors and matrices.[7] All advanced works in mathematical programming make extensive use of these mathematical tools. An introductory discussion of matrices is given in Chapter 9, but in the context of input-output analysis rather than linear programming.

Problems

5-1. Solve Problem 3-1 by the simplex method.

5-2. Solve Problem 3-2 by the simplex method.

5-3. Assume that the manufacturer in Problem 3-3 can also produce and sell a third product (C). Each unit of C produced by the manufacturer requires the following operations: 1 hour of grinding, 4 hours of turning, 4 hours of assembly, and 6 hours of testing. Each unit of C contributes \$2.50 to profit and overhead. Use the simplex method to determine the optimum amount of A, B, and C to produce.

[7] An introductory exposition of the simplex method in terms of vectors and matrices can be found in *Microeconomics and Decision Models of the Firm* by Naylor and Vernon. A more advanced exposition is given in G. Hadley, *Linear Programming*.

5-4. A job shop manufacturer has in process six jobs that must be completed during the following week. The number of hours required to complete each job in each of the manufacturer's three shops is indicated by the table below. A job may be "split" and completed in one or more of the three shops. The operating cost per hour for Shops #1, #2, and #3 are \$100, \$110, and \$115 respectively. Each of these shops may be operated for a maximum of 40 hours per week. Determine an optimum production schedule for the job shop by the simplex method.

Shop	*Hours Required Per Job*					
	1	2	3	4	5	6
1	2	13	8	3	7	12
2	1	15	9	4	6	10
3	3	12	10	3	5	14

5-5. Use the simplex method to determine the non-negative values of X_1 and X_2 that maximize $Z = 5X_1 + 3X_2$ subject to

$$X_2 \geq \frac{9}{5},$$

$$2X_1 + X_2 \leq 5,$$

$$3X_1 + 4X_2 \leq 12.$$

5-6. Solve the Johnson Manufacturing Company problem by the simplex method (Problem 3-5).

5-7. In Problem 5-6, assume that there is now a *variable* cost associated with the production of each widget. Thus, when the Johnson Company uses the process that requires five labor units and one machine unit, the variable cost per widget is 1.0; when the process that requires one labor unit and two machine units is employed, the variable cost per widget is 2.0. (An example of variable cost might be electric energy consumed.) Derive the marginal cost curve for widgets by calculating the minimum addition to total variable cost for each additional widget as the output of widgets increases from zero to capacity.

5-8. Ace Advertising Agency has estimated an "effective exposure" measure for each of two advertising media in behalf of its client, the Boyles Broom Company. Interestingly, Ace estimates that each full page color advertisement in Playgirl magazine has the same effective exposure of 10,000 units as the sponsorship of a half hour television show. The Boyles Broom Company has a \$100,000 advertising budget for the next six months and stipulates that its advertising program must include full-page ads in at least two issues of Playgirl. (No more than one full page per issue is worthwhile, and Playgirl is a monthly magazine.) If it costs \$20,000 per television show and \$10,000 for a page in Playgirl, how should Boyles allocate its budget for the next six months in order to maximize total effective exposure?

5-9. Suppose the Boyles Broom Company of Problem 5-8 should also advise Ace that it wished to sponsor at least three television shows in addition to the two pages in Playgirl. How does this new stipulation affect the advertising program determined in Problem 5-8? How should Ace advise Boyles concerning this new requirement?

Duality and Sensitivity Analysis 6

This chapter will be concerned with extensions of linear programming. First, we shall discuss the dual problem and interpret its economic significance. Then, we shall briefly illustrate sensitivity analysis and parametric programming. Sensitivity analysis refers to an investigation of the effect of changes in the coefficients and constants on the optimum solution.

THE DUAL PROBLEM

Every linear programming problem has a *dual problem* associated with it. For example, in the television assembly problem the method of solution chosen determines the production of that quantity of black-and-white and color sets which maximizes the returns from the sets and, at the same time, does not violate the production capacity restrictions. The dual to this problem determines that value or price assigned to each unit of process which minimizes the value of the total input or processes used while not violating, as a lower limit, the contribution to profit and overhead per unit for each type of set. The solution of one of these problems also solves the other.

The *primal problem* of the television manufacturer is one of determining Q_B, $Q_C \geq 0$, which will maximize $Z = 15Q_B + 45Q_C$ subject to

$$1.0Q_C \leq 50\text{—color tubes} \tag{1}$$

$$1.0Q_B + 1.6Q_C \leq 240\text{—set assembly} \tag{2}$$

$$0.5Q_B + 2.0Q_C \leq 162\text{—final inspection} \tag{3}$$

This problem is described in tabular form in Table 6-1. The restrictions may be read from this table by interpreting a single vertical line as + and the double vertical line as $\leq$. The function to be maximized is given in the bottom row, namely $15Q_B + 45Q_C$.

TABLE 6-1

	+	$\leq$
	Q_B	Q_C	
	0	1.0	50
	1.0	1.6	240
	0.5	2.0	162
Max	15	45	

To obtain the dual problem, we must extend Table 6-1 to the form of Table 6-2. Reading down each column of Table 6-2 the restrictions may be derived by interpreting a single horizontal line as + and the double horizontal line as $\geq$. The function to be minimized is given by the column to the right of the vertical double line, namely $50N_1 + 240N_2 + 162N_3$.

TABLE 6-2

		+	$\leq$	
		Q_B	Q_C	*Min*
	N_1	0	1.0	50
+	N_2	1.0	1.6	240
+	N_3	0.5	2.0	162
$\geq$	*Max*	15	45	

The dual problem of the television manufacturer is one of determining N_1, N_2, $N_3 \geq 0$ which will minimize $g = 50N_1 + 240N_2 + 162N_3$ subject to

$$1.0N_2 + 0.5N_3 \geq 15\text{—unit profit of black-and-white sets;} \quad (4)$$

$$1.0N_1 + 1.6N_2 + 2.0N_3 \geq 45\text{—unit profit of color sets.} \quad (5)$$

In this case N_1, N_2, and N_3 are the values, or prices, assigned respectively to each color picture tube, each man-hour of set assembly-line capacity, and each man-hour of final testing capacity.

At this point, it may help you to refer back to the section in Chapter 5 entitled *The Value of a Marginal Unit of Production Capacity*. There, we referred to the variables N_1, N_2, and N_3 as shadow prices and interpreted them as opportunity profits of the resource restraints.

The objective of the dual problem is to minimize g, the value of the total input or processes used, (color picture tubes, set assembly-line capacity, and final testing capacity), without violating the contribution to profit and overhead per unit for black-and-white and color sets as a lower limit. The profit restraint for black-and-white sets means that one man-hour of set assembly time multiplied by its value per man-hour plus 0.5 man-hours of final testing time multiplied by its value per man-hour must exceed the unit profit contribution of black-and-white sets. The color-set profit restraint means that 1.0 color picture tube multiplied by its value per picture tube, plus 1.6 man-hours of set assembly time multiplied by its value per man-hour, plus 2.0 man-hours of final testing time multiplied by its value per man-hour must be greater than or equal to the per unit contribution to profit of color sets.

Before solving the dual problem we will state *Duality Theorem 1*. This remarkable theorem should give you some insight into the reason for our interest in the dual problem. So far, we have merely pointed out that by reversing everything in the primal problem we obtain something called the dual problem.

> *Duality Theorem 1*. If either the primal or the dual problem has a finite optimum solution, then the other problem has a finite optimum solution, and the values of the two objective functions are equal.[1]

Thus, we discover that maximizing Z subject to the restraints (1)–(3) yields the same value as minimizing g subject to the dual-problem restraints (4)–(5). In other

[1] A proof of this theorem, using matrix algebra, can be found in Saul Gass, *Linear Programming: Methods and Applications* (New York: McGraw-Hill Book Co., 1964), p. 84.

words, since the maximum value of Z in Chapter 5 was found to be \$4335, the minimum value of g must also equal \$4335.

Using the simplex method, we now solve the dual problem. We must introduce slack and artificial variables to obtain the initial basic feasible solution. Table 6-3 is the initial simplex table. Note that we use the letter T for the slack variables and A for the artificial variables.

TABLE 6-3

		Column	1	2	3	4	5	6	7	
		C_j	50	240	162	0	M	0	M	
w	C_i	Solution	N_1	N_2	N_3	T_1	A_1	T_2	A_2	P_0
	M	A_1	0	1.0	0.5	−1	1	0	0	15
	M	A_2	1.0	1.6	2.0	0	0	−1	1	45
		Z_j	M	$2.6M$	$2.5M$	$-M$	M	$-M$	M	$60M$
		$C_j - Z_j$	$50 - M$	$240 - 2.6M$	$162 - 2.5M$	M	0	M	0	

To save space we shall show the final simplex table only. Table 6-4 represents the optimum solution. Note that we have also dropped the artificial variables, since their only function was to provide an initial solution.

TABLE 6-4

		Column	1	2	3	4	5	
		C_j	50	240	162	0	0	
Row	C_i	Solution	N_1	N_2	N_3	T_1	T_2	P_0
1	240	N_2	$5/12$	1	0	$-5/3$	$5/12$	6.25
2	162	N_3	$5/6$	0	1	$4/3$	$-5/6$	17.50
		Z_j	35	240	162	−184	−35	4335
		$C_j - Z_j$	15	0	0	184	35	

The minimum value of g in the dual problem is indeed \$4335. There are other aspects of the optimum solution to the dual that should be related to the optimum solution to the primal problem (Table 5-4). For example, the optimum values of Q_B

and Q_C appear as the $C_j - Z_j$ values in columns 4 and 5 of the dual table, and the optimum values of N_1, N_2, and N_3 appear in the $C_j - Z_j$ row of the primal table, Columns 3, 4, and 5.

The economic interpretation of the primal-dual relationship can be best explained by introducing a second duality theorem. First, however, recall that the primal variables of our television problem are Q_B and Q_C, the quantities of the two types of sets which are produced. The dual slack variables, T_1 and T_2, can be viewed as related to Q_B and Q_C, respectively, via the dual restraints. That is, T_1 measures the excess of value imputed to the resources producing Q_B, above the unit profit contribution of Q_B. Rearranging restraint (4) after adding the slack variable gives

$$T_1 = 1.0N_2 + 0.5N_3 - 15. \tag{6}$$

Furthermore, the dual variables, N_1, N_2, and N_3, are related to the primal slack variables S_1, S_2, and S_3 via the primal restraints. For example, the primal slack variable S_1 measures the quantity of color picture tubes *not* used in the production of television sets. The variable N_1 is, of course, the imputed value, or price, of color picture tubes. Rearranging restraint (1) after adding the slack variable gives

$$S_1 = 50 - 1.0Q_C. \tag{7}$$

Duality Theorem 2. (a) If a dual variable (N) is optimally nonzero, then its related primal slack variable (S) must optimally be zero.

(b) If a dual slack variable (T) is optimally nonzero, then its related primal variable Q must optimally be zero.[2]

[2] The proof can be sketched as follows: Add slack variables to restraints (1)–(5). Multiply (1) through by N_1, (2) by N_2, and (3) by N_3. Add these three equations. The right-hand side is $50N_1 + 240N_2 + 162N_3$, or g. Hence, the left-hand side is equal to g. Thus,

$$g = 1.0Q_CN_1 + S_1N_1 + 1.0Q_BN_2 + 1.6Q_CN_2 + S_2N_2 + 0.5Q_BN_3 + 2.0Q_CN_3 + S_3N_3.$$

Now multiply (4) by Q_B and (5) by Q_C. Add these two equations. The right-hand side is $15Q_B + 45Q_C$, or Z. Thus,

$$Z = 1.0N_2Q_B + 0.5N_3Q_B - T_1Q_B + 1.0N_1Q_C + 1.6N_2Q_C + 2.0N_3Q_C - T_2Q_C.$$

Duality Theorem 1 states that the optimal values of g and Z must be equal. Hence, equating the right-hand sides of g and Z and dropping terms appearing on both sides, we obtain:

$$S_1N_1 + S_2N_2 + S_3N_3 = -T_1Q_B - T_2Q_C,$$

or

$$T_1Q_B + T_2Q_C + S_1N_1 + S_2N_2 + S_3N_3 = 0.$$

At least one variable in each term must be zero to satisfy the equation, which is what Duality Theorem 2 states.

Since the selection of the maximizing problem as the primal problem was arbitrary (the minimizing problem could just as well have been called the primal problem), the theorem will also hold true if we reverse the names. That is, let the minimizing problem be the dual. Then, for example, the first part of the theorem states that if Q is optimally nonzero, then its related T must optimally be zero.

In short, Duality Theorem 2 can be stated as the following "either-or" conditions:

$$Q_B T_1 = 0, \tag{8}$$

$$Q_C T_2 = 0, \tag{9}$$

$$N_1 S_1 = 0, \tag{10}$$

$$N_2 S_2 = 0, \tag{11}$$

$$N_3 S_3 = 0, \tag{12}$$

where the conditions hold true only for the values of the variables in the optimum solution. Thus, in (8) *either* Q_B is zero *or* T_1 is zero.

We can demonstrate the validity of the "either-or" conditions for the television problem, where the solutions were:

Primal solution	*Dual solution*
$Q_B = 184,$	$T_1 = 0,$
$Q_C = 35,$	$T_2 = 0,$
$S_1 = 15,$	$N_1 = 0,$
$S_2 = 0,$	$N_2 = 6.25;$
$S_3 = 0,$	$N_3 = 17.50.$

Hence,

$$Q_B T_1 = (184)(0) = 0, \tag{8a}$$

$$Q_C T_2 = (35)(0) = 0, \tag{9a}$$

$$N_1 S_1 = (0)(15) = 0, \tag{10a}$$

$$N_2 S_2 = (6.25)(0) = 0, \tag{11a}$$

$$N_3 S_3 = (17.50)(0) = 0. \tag{12a}$$

Now we shall examine the economic sense of these conditions. If Q_B, the optimal number of black-and-white sets, is nonzero, then $T_1 = 0$ means that the imputed value of resources used in producing a black-and-white set exactly equals the unit profit contribution of black-and-white sets. This is also true of color sets, and it means that we have selected the correct prices to exactly apportion profits to the resources. But what is the meaning of the requirement that Q_B must equal zero if T_1 is positive? If T_1 is positive, the value of the resources used in producing a black-and-white set *exceeds* the unit profit contribution of a black-and-white set. It is clear that Q_B *should* be zero in this case. To illustrate, we refer to the dual restraint (4), which, after adding the slack variable, becomes

$$1.0N_2 + 0.5N_3 - T_1 = 15. \tag{13}$$

The sum of the first two terms on the left is the value of the resources used in producing a black-and-white set. The constant, 15, is the unit contribution to profit of the set. A positive T_1 means that

$$1.0N_2 + 0.5N_3 > 15, \tag{14}$$

which is the signal that Q_B should, therefore, be zero.[3] An analogous argument can be made for condition (9).

"Either-or" conditions (10), (11), and (12), are similar; hence, we will limit our discussion to the rationale for (10a) and (11a). In (10a), since $S_1 = 15$, then N_1 must be zero. But what is S_1? This is a slack variable which indicates that, in the optimum solution, there are 15 *unused* color television tubes. If there are surplus color tubes then it is only common sense to expect that the imputed value (opportunity profit) of a color tube must be zero. An additional color tube would not be of any value, since it is the *other* resources that are binding, and the extra color tube could only increase the existing surplus of tubes.

The other side of the condition is illustrated by (11a). Here, since $N_2 = 6.25$, then S_2 must be zero. There could be no surplus set assembly man-hours (S_2 could not be positive) or else the value of those man-hours would necessarily be zero. Since S_2 is zero, this means that all set assembly man-hours are being used, and that this restraint is binding. Since an additional assembly man-hour would permit an increase in profit of 6.25, its imputed value is positive.

Thus, Duality Theorem 2 specifies a set of "either-or" conditions which relate the primal and dual variables in the optimum solution. The conditions have very plausible economic interpretations.

To conclude the discussion of duality, we should point out that sometimes a computational advantage can be gained by solving the dual problem, rather than the

[3] It is not possible for T_1 to be *negative*, since all variables are restricted to non-negative values in linear programming.

primal one. This will be true if the dual problem has fewer restraints than the primal, since the number of simplex tables depends more upon the number of restraints than upon the number of variables.

SENSITIVITY ANALYSIS

In industrial applications of linear programming, the estimates of the coefficients of the objective function, the coefficients in the restraints and the constants in the restraints are seldom known with complete certainty. In many problems the uncertainty is great; therefore, the question of the effect of bad estimates is important. The object of sensitivity analysis is to determine the effect of changes in the coefficients on the optimal solution.

We now return to the graphical analysis in order to illustrate the type of information that sensitivity analysis can provide. The purpose of this discussion is mainly to emphasize the importance of sensitivity analysis; we shall conclude this section with an elementary exposition of a more systematic and powerful procedure, known as *parametric programming*.[4]

As our example we shall consider the sensitivity of the optimum solution to the profit function coefficients. The graphical solution to the television assembly problem is shown in Figure 6-1.

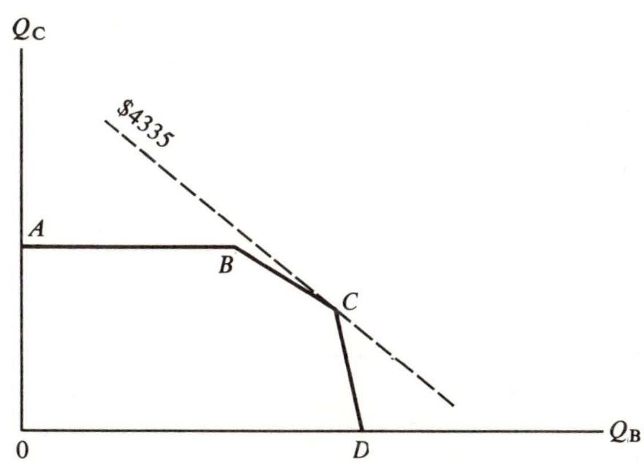

Figure 6-1. Graphical Solution.

[4] For a good exposition, see F. Hillier and G. Lieberman, *Introduction to Operations Research* (San Francisco: Holden-Day, 1967), p. 499.

Point C is the optimum solution, since it touches the highest iso-profit line (\$4335). The optimum point is the intersection of two restraint lines, namely BC and CD. The restraint BC is given in equation form as

$$Q_C = -0.25Q_B + 81, \tag{15}$$

and the restraint CD is given by

$$Q_C = -0.625Q_B + 150. \tag{16}$$

The equations were transformed into this form for easy comparison of the slopes of the lines. The profit function is

$$Q_C = -0.333Q_B + \frac{Z}{45}. \tag{17}$$

Now we shall extend the restraint lines, as shown in Figure 6-2.

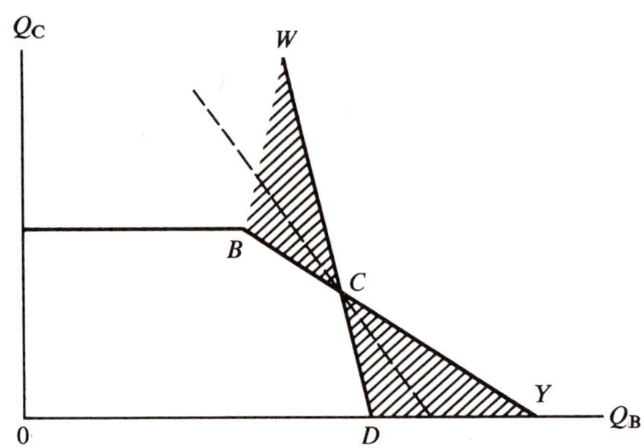

Figure 6-2. Graphical Solution Indicating Range of Profit Function Variation with no Change in Solution.

An examination of Figure 6-2 reveals that the slope of the dashed profit function can vary anywhere between the slope of BCY and the slope of WCD without altering the optimum solution. This is equivalent to the statement that the dashed line can rotate about C anywhere within the shaded region without changing the optimum solution.

For the current solution, this criterion can be demonstrated easily:

$$\text{slope of } BCY > \text{slope of profit function} > \text{slope of } WCD.$$

Or, from equations (15)–(17)

$$-0.25 > -0.333 > -0.625. \qquad (18)$$

Let the profit function be written with unknown coefficients:

$$Z = P_B Q_B + P_C Q_C.$$

Then, equation (17) becomes

$$Q_C = -\frac{P_B}{P_C} Q_B + \frac{Z}{P_C}, \qquad (19)$$

and inequality (18) is now

$$-0.25 > -\frac{P_B}{P_C} > -0.625.$$

Or, multiplying through by -1,

$$0.25 < \frac{P_B}{P_C} < 0.625. \qquad (20)$$

In other words, the ratio of P_B to P_C can vary anywhere between 0.25 and 0.625, while the optimum solution (point C) will remain unaffected. For example, let P_C be fixed at the current value of \$45. How sensitive is the optimum solution to changes in P_B, the unit profit of black-and-white sets?

From inequality (20), it follows that

$$11.25 < P_B < 28.125.$$

That is, for P_C fixed at \$45, the unit profit of black-and-white sets can vary anywhere between \$11.25 and \$28.125 without changing the optimum solution.

We can also construct a graphical region of values of P_B and P_C which are consistent with the optimum solution at point C. From (20) we can write two inequalities:

$$P_B > 0.25 P_C$$

and

$$P_B < 0.625 P_C.$$

These define the shaded region in Figure 6-3.

Sensitivity analysis, by identifying those coefficients to which the solution is most sensitive, permits management to allocate funds toward improving estimation in an efficient manner. Thus, if the solution is relatively insensitive to a particular coefficient, it is unnecessary to spend large sums of money in an attempt to get better estimates of that coefficient.

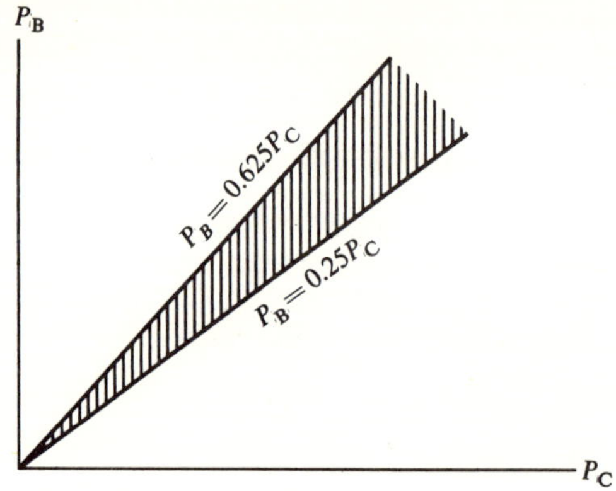

Figure 6-3. Region of Values of P_B and P_C Consistent with Optimum Solution.

Parametric Programming

We shall now consider a variant of sensitivity analysis—*parametric programming*. We differentiate parametric programming from sensitivity analysis by the range of values over which a parameter is varied. In parametric programming, we study the effect of a parameter on the objective function over a much greater range of values than we do in sensitivity analysis.

Suppose that a firm seeks to maximize the contribution to profit and overhead, Z, where

$$Z = 1.5X_1 + 1X_2. \tag{21}$$

The two variables X_1 and X_2 denote the quantities of the firm's two products. The firm is constrained by two capacity restraints:

$$2X_1 + 1X_2 \leq 8 \tag{22}$$

and

$$\frac{1}{3}X_1 + 1X_2 \leq 3 \tag{23}$$

Of course, negative production is prohibited. Thus,

$$X_1 \geq 0, \qquad X_2 \geq 0. \tag{24}$$

Figure 6-4 shows the graphical solution to the problem. We see that Z is a maximum at point T, where $X_1 = 3$ and $X_2 = 2$. The maximum value of Z is

$$Z = 1.5(3) + 1(2) = 6.5.$$

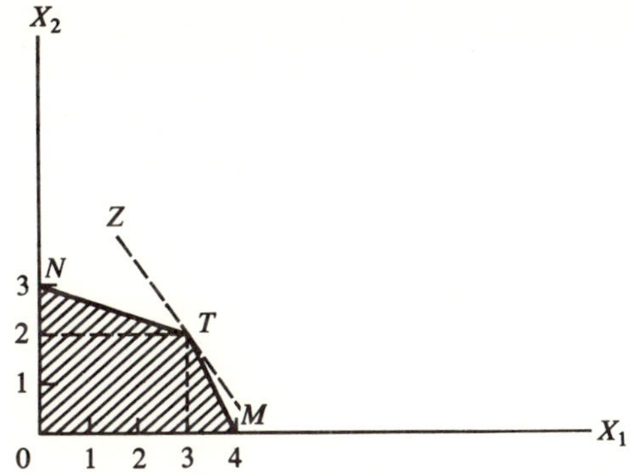

Figure 6-4. Graphical Solution.

Thus far, we have merely solved a very simple linear programming problem. Now we wish to change the problem by creating a new restraint—"working capital," or K. We assume that every unit of X_1 and X_2 that the firm produces requires a unit of K. Furthermore, the firm does not have any K on hand, but it can obtain any quantity of K that it wishes from the bank at a price of 1.0 per unit. In effect, the new restraint can be written as

$$1X_1 + 1X_2 \leq K.$$

Having solved the problem prior to the addition of the K restraint, we know that at point T the firm earns a profit of 6.5 and produces 3 units of X_1 and 2 units of X_2. The working-capital restraint would force the firm to buy 5 units of K at a total cost of 5.0. Thus, at point T the firm would still make a positive contribution to profit of $6.5 - 5.0 = 1.5$. However, this is no longer the optimum.

In order to find the optimum, we need to construct a demand schedule for K. For our simple example, this can be done easily. More complex problems demand a knowledge of parametric programming algorithms, which are beyond the scope of this book.

In principle, the construction of a demand schedule is quite straightforward. First, add the new restraint to the other two restraints:

$$2X_1 + 1X_2 \leq 8,$$

$$\frac{1}{3}X_1 + 1X_2 \leq 3,$$

$$1X_1 + 1X_2 \leq K.$$

Next, vary the unknown K from $K = 0$ to an appropriately large value of K (in our example, $K = 5$ is appropriate). For each value of K, solve the linear programming problem by the simplex method and obtain the imputed value, or shadow price, of K. Finally, plot the resulting values as shown in Figure 6-5.

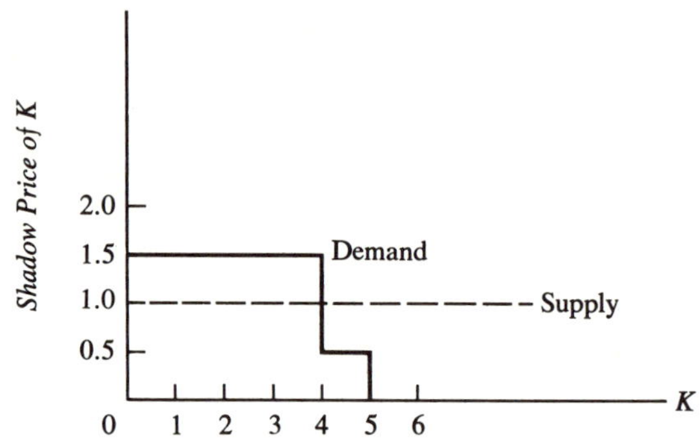

Figure 6-5. Demand and Supply for K.

The dashed line in Figure 6-5 represents the supply price of K. It is clear that the firm should *not* buy 5 units of K, as would be required at Point T. The demand for K in the figure shows that the fifth unit of K would add only 0.5 to profit, whereas the cost of obtaining the fifth unit is 1.0. The firm should buy only 4 units of K.

Returning to Figure 6-4, we now trace through the reasoning involved in deriving the demand for K. Begin at the origin and envisage the K restraint for $K = 1.0$. The feasible region is now simply a triangle, where the sides on the axes each extend from the origin to 1.0. The optimum production plan must be at one of the corner points, $X_1 = 1, X_2 = 0$ or $X_1 = 0, X_2 = 1$. The objective function tells us which one. Since the coefficient of X_1 is 1.5, compared to only 1.0 for X_2, the maximum profit will be attained at $X_1 = 1, X_2 = 0$.

Likewise, as K is increased from 1 to 4, one can imagine the restraint line shifting outward from the origin. Hence, the point representing the optimum production plan will also shift outward along the X_1 axis. The optimum production plan changes from $X_1 = 1$ to $X_1 = 4$, with X_2 always zero. Therefore, over this range, the shadow price of K is the same as the coefficient of X_1 in the profit function.

Next, consider what happens to profit as the value of K changes from 4 to 5. This is shown graphically in Figures 6-6a and 6-6b. The optimum point is M in Figure 6-6a ($K = 4$) and T in Figure 6-6b ($K = 5$).

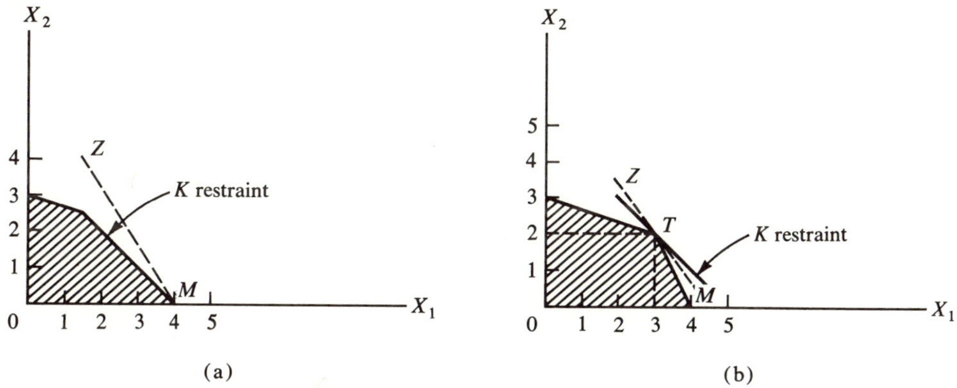

Figure 6-6

The change in profit as a result of the increase in K from 4 to 5 (this is the definition of the shadow price of K) is calculated below:

$Z = 1.5(3) + 1(2) = 6.5$	——	point T $(K - 5)$
Less $Z = 1.5(4) + 1(0) = 6.0$	——	point M $(K = 4)$
Shadow price of fifth unit of K: 0.5		

Since the cost of obtaining the fifth unit of K is 1.0, it is not optimal to add the fifth unit.

Thus, parametric programming is a very useful tool for management. Not only is the optimal program itself determined, but management also gains information about the value of inputs over wide ranges.

Problems

6-1. Show that by formulating the dual of the dual problem, the primal problem is obtained. Take the dual of the television assembly problem as your example.

6-2. Problem 3-1 is a maximizing problem in which the constants in the restraints, 12 and 8, represent resource availabilities. Suppose management is uncertain of its estimates of these constants and wishes to know how sensitive the solution is to these estimates. Perform a graphical sensitivity analysis similar to the one described in the text. *Hint:* Analyze the dual problem.

6-3. Suppose that the Johnson Manufacturing Company of Problem 5-6 has the opportunity to buy additional labor units at 0.1 per unit. Should it do so? Does this hold true for any quantity of additional labor? Explain.

6-4. Derive a demand schedule for K for the following problem:

Maximize $$Z = 1.0X_1 + 1.5X_2$$

subject to

$$2X_1 + 1X_2 \leq 8,$$

$$\frac{1}{3}X_1 + 1X_2 \leq 3,$$

$$1X_1 + 1X_2 \leq K,$$

and

$$X_1 \geq 0, \qquad X_2 \geq 0.$$

6-5. Determine the optimum production plan for Problem 6-4 given that the firm must pay a price of 1.0 per unit of K. Suppose the price of K drops to 0.5; how does this change affect the optimal production plan?

7 Special Methods: Transportation and Assignment

Having completed the explanation of the general method for solving linear programming problems, we shall now examine two special methods that are useful for solving particular types of problems. The first of these special methods is the transportation method. It is used to solve that class of problems in which (1) the sum of resources equals the sum of requirements, both of which are stated in only one kind of unit; and (2) all of the exchange coefficients are either zero or one. The assignment method is applicable to that class of problems in which n jobs must be assigned to n facilities having different performance characteristics. These techniques are presented because they offer significant computational efficiency over the simplex method, not because they solve problems the simplex method cannot solve.

THE TRANSPORTATION METHOD

The transportation method was developed to solve the problem of minimizing the cost of distributing a product from a number of sources or origins to a number of destinations.[1] The use of the transportation method is not restricted to transportation problems alone; it is applied to that class of linear programming problems in which the requirements and resources are stated in terms of only one kind of unit, such as product, boxcars, or trucks. The transportation method involves the following steps:

1. Define the problem, including the objective function to be optimized and the restraints imposed on the solution (including the slack variable);
2. Develop an initial solution;
3. Evaluate alternatives to the existing solution;
4. Change the solution;
5. Repeat steps 3 and 4 until an optimum solution is found.

The transportation method of solving linear programming problems is most easily explained by reference to a simple problem. Consider a manufacturer who operates 2 factories and ships his product to 3 different destinations. The table below indicates the capacities of the 2 factories (F_1, F_2) and the quantity of product that is required at each of 3 selling points, or destinations (D_1, D_2, D_3,).

TABLE 7-1

Factory	*Capacity*	*Destination*	*Requirements*
F_1	150	D_1	90
F_2	100	D_2	80
Total	250	D_3	60
		Total	230

The cost of shipping 1 unit of the product from each of the 2 factories to each of the 3 destinations is given by the following cost matrix:

	Destination		
Factory	D_1	D_2	D_3
F_1	0.29	0.35	0.25
F_2	0.24	0.26	0.33

With this information we can construct Table 7-2.

[1] F. L. Hitchcock, "The Distribution of a Product from Several Sources to Numerous Localities," *J. Math. Phys.*, 20 (1941), 224–230.

TABLE 7-2

Factory	Destination D_1	D_2	D_3	S_L	Factory Capacities
F_1	0.29 / X_{11}	0.35 / X_{12}	0.25 / X_{13}	0 / X_{1S}	150
F_2	0.24 / X_{21}	0.26 / X_{22}	0.33 / X_{23}	0 / X_{2S}	100
Destination Requirements	90	80	60	20	250

It will be noted that in addition to the 3 destinations D_1, D_2, and D_3, another "destination" is shown. This destination is called slack, and it is added because the total requirements must equal the total resources before the transportation method can be used. Table 7-1 indicates that total requirements are 230 units, whereas total resources (factory capacities) are 250 units. The slack variable S_L in Table 7-2 is used to equate requirements and resources and is equal to 20 units. Thus, the final solution must contain 20 units of unused capacity; that is, there will be a total of 20 units that will not be shipped to any destination. Obviously, the cost of not shipping a unit is zero. The quantities to be shipped are found in the lower right hand corner of each cell and are denoted by X_{11}, X_{12}, etc. The transportation costs appear in the upper left of each cell. The last column in the table and the last row represent the factory capacities and destination requirements, respectively. There are 2 factories and 4 destinations. The quantities shown in this column and row are called rim requirements. The number of rim requirements totals 6.

The Initial Solution

Having completed Step 1 we must now determine an initial solution. This can be accomplished in a variety of ways. We can develop a solution by inspection, we can use the present program if one exists, or we can use what is known as the "northwest-corner" solution, presented in Table 7-3.

TABLE 7-3

Factory	Destination D_1	D_2	D_3	S_L	Factory Capacities
F_1	0.29 / 90	0.35 / 60	0.25	0	150
F_2	0.24	0.26 / 20	0.33 / 60	0 / 20	100
Destination Requirements	90	80	60	20	250

The following rule is known as the northwest-corner rule. Beginning with the northwest corner (cell F_1D_1), assign quantities to cells in that row until the destination requirements or factory capacities are fully used. Then go to the next row and assign quantities until the capacity is fully used. In our problem we assign 90 units of factory F_1 to destination D_1. This satisfies the requirements of D_1 and still leaves 60 units of F_1 capacity unused. This is then applied to the requirements of D_2. But D_2 requires 80 units, so we move down to the next row (F_2) for the additional 20 units. The remaining 80 units of F_2 capacity are assigned to D_3 (60 units) and S_L (20 units). The total cost of the initial program is

$$90(0.29) + 60(0.35) + 20(0.26) + 60(0.33) + 20(0) = \$72.10.$$

It should be noted that by assigning 20 units of slack to F_2 we have determined a production schedule for the 2 factories. F_1 will operate at full capacity at this point, whereas F_2 will operate at 80% capacity. This completes Step (2) of the method.

Evaluation of Alternatives

The next problem we face is that of deciding whether or not the solution we have found is optimum. In other words, is there any other program that, when put into operation, will result in a decrease in total cost? To answer this question we must evaluate the alternatives available to us. In this case, the alternatives are the cells that are not presently being used. We want to know the effect on cost of adding 1 unit to an unused cell. Let us evaluate cell F_2D_1 as an illustration. Adding 1 unit to F_2D_1 will add \$0.24 to cost. But since destination requirements must be satisfied we will subtract 1 unit from F_1D_1. This will decrease cost by \$0.29. Now, however, the factory capacity restrictions are not satisfied, F_1 being 149 units, whereas F_2 is 101 units. To correct this, we can add 1 unit to F_1D_2, which increases cost by \$0.35, and subtract 1 unit from F_2D_2, which decreases cost by \$0.26. The total effect on cost is

$$(+0.24 - 0.29 + 0.35 - 0.26 = +\$0.04)$$

by adding and subtracting a unit of product on the route

$$(F_2D_1 \rightarrow F_1D_1 \rightarrow F_1D_2 \rightarrow F_2D_2).$$

The route just described can be called a "closed-path" evaluation. A closed-path evaluation, therefore, is one in which the evaluation procedure begins with an unused cell and, by a series of horizontal and vertical moves, passes through occupied cells until the original starting point is reached. If we cannot trace such a path in a table, we have what may be called an "open path." This distinction will prove useful later when we come to the matter of degeneracy in the transportation method. For the moment we can note that in evaluating unused cells we can use only closed paths.

We can now apply an interpretation to the fact that cell F_2D_1 has been evaluated as $+\$0.04$. At this point in the solution shipping product from F_2 to D_1 will add \$0.04 per unit to total cost. It is apparent then that any unused cell will not be considered for inclusion in a new solution if the change in cost is positive (that is, if the cell is evaluated unfavorably). If a cell is evaluated favorably, so that the change in cost is negative, the cell will be considered for inclusion in a new solution. The exact cell that will be brought in is determined by the magnitude of the evaluation; the one with the most negative value will be introduced.

We shall present the evaluation of one more cell to further demonstrate the procedure. Let us evaluate F_1D_3. Adding 1 unit to F_1D_3 will add \$0.25 to total cost. But once again we must consider the rim requirements. Therefore, we shall subtract a unit from F_2D_3 (decreasing cost by \$0.33), add a unit to F_2D_2 (increasing cost by \$0.26), and subtract a unit from F_1D_2 (decreasing cost by \$0.35). The net effect on cost of using F_1D_3 by the route $F_1D_3 \rightarrow F_2D_3 \rightarrow F_2D_2 \rightarrow F_1D_2$ is

$$+0.25 - 0.33 + 0.26 - 0.35 = -\$0.17.$$

In other words, at this stage of the solution, adjustments similar to those described above will reduce total cost by \$0.17 per unit shipped from F_1 to D_3. The evaluation of the remaining unused cell F_1S_L is

$$+0 - 0 + 0.26 - 0.35 = -0.09.$$

Since F_1D_3 has been evaluated as $-\$0.17$ it will improve the present solution at a rate of \$0.17 per unit as compared to \$0.09 per unit for F_1S_L. It is on this basis that we choose to include F_1D_3 in a new solution.[2] The problem now is to determine how much product can be shipped from F_1 to D_3 and what changes are required in the present solution.

Changing the Solution

Our evaluation procedure for cell F_1D_3 involved adding 1 unit each to F_1D_3 and F_2D_2 and subtracting 1 unit each from F_1D_2 and F_2D_3. We can continue these adjustments so long as we satisfy the rim requirements. Referring to Table 7-3, we see that 60 units can be added to F_1D_3 by subtracting 60 units each from F_2D_3 and F_1D_2 and adding 60 units to F_2D_2. The new program is shown in Table 7-4. The total cost of the new program is

$$90(0.29) + 60(0.25) + 80(0.26) + 0(20) = \$61.90.$$

TABLE 7-4

Factory	*Destination* D_1	D_2	D_3	S_L	*Factory Capacities*
F_1	0.29 / 90	0.35	0.25 / 60	0	150
F_2	0.24	0.26 / 80	0.33	0 / 20	100
Destination Requirements	90	80	60	20	250

[2] The reader should be aware that the method is equally successful in solving problems of maximization of profits. In that case, the elements of the matrix would be profit per unit, and a cell would be favorably evaluated if the evaluation procedure yielded a positive value. Problems can also be stated in terms of time, distance, etc.

We have now completed the first four steps of the transportation method. An iterative process consisting of Steps 3, 4, and 5 will now be used until a final solution is found.

Once again it is necessary to evaluate the unused cells in order to determine whether or not we can improve the new program we have developed. Before we can do this we must introduce the notion of degeneracy. Degeneracy occurs in the transportation method when the number of rim requirements minus 1 is greater than the number of occupied cells. In our case rim requirements minus 1 equals 5, whereas the number of occupied cells numbers only 4. In other words, we have the situation mentioned earlier. For example, cell F_1S_L must be evaluated, but we cannot do this following a closed path. Every path we try turns out to be open, because the occupied cells in the same row as F_1S_L do not have a corresponding occupied column cell. To resolve this problem we shall make use of an artificial quantity that we shall denote by the Greek letter Δ (delta). If Δ is placed in the lower right-hand corner of a cell, that cell is considered occupied by a very small quantity. The quantity is so small that it does not affect the capacity restrictions or destination requirements. Nor will it have any significance if it should appear in the final solution. Its use and significance are confined solely to the evaluation of unused cells. Once a Δ has been inserted into the solution it will remain there until it is removed by subtraction or a final solution is reached. The use of Δ is illustrated by reference to Table 7-5.

You will notice that Δ has been placed in cell F_2D_1, although it could have

TABLE 7-5

Factory	*Destination* D_1	D_2	D_3	S_L	*Factory Capacities*
F_1	0.29 / 90	0.35 / +0.04	0.25 / 60	0 / −0.05	150
F_2	0.24 / Δ	0.26 / 80	0.33 / +0.13	0 / 20	100
Destination Requirements	90	80	60	20	250

been placed in any one of the other unused cells: F_1D_2, F_1S_L, or F_2D_3. This permits evaluation of the unused cells. The evaluations of the three alternatives are shown in the left-hand corner of the unused cells, and they have been calculated as

$$F_1D_2 = +0.35 - 0.29 + 0.24 - 0.26 = +0.04,$$
$$F_1S_L = +0 - 0.29 + 0.24 - 0 = -0.05,$$
$$F_2D_3 = +0.33 - 0.25 + 0.29 - 0.24 = +0.13.$$

The cell F_1S_L has been favorably evaluated. This means that it is desirable at this point to introduce slack (unused capacity) for Factory 1. How much slack should we introduce? Again, keeping in mind the necessity for satisfying rim requirements and using our evaluation path ($F_1S_L \rightarrow F_1D_1 \rightarrow F_2D_1 \rightarrow F_2S_L$), we can determine the required amount of slack to assign to F_1. We can add 20 units of slack to F_1 by subtracting 20 units from F_2. By so doing we violate the capacity restriction of F_1 so we must reduce the amount shipped from F_1 to D_1 by 20 units. This in turn upsets the destination requirements of D_1. We correct this by adding 20 units of product to F_2D_1. This completely satisfies the rim requirements. Table 7-6 shows the new program.

The cost of the above program is

$$70(0.29) + 60(0.25) + 20(0) + (20 + \Delta)(0.24) + 80(0.26) = \$60.90.$$

The evaluation of alternatives is shown in the left-hand corner of the unused cells.

TABLE 7-6

Factory	Destination D_1	D_2	D_3	S_L	Factory Capacities
F_1	0.29 / 70	0.35 / +0.04	0.25 / 60	0 / 20	150
F_2	0.24 / $20 + \Delta$	0.26 / 80	0.33 / +0.13	0 / +0.05	100
Destination Requirements	90	80	60	20	250

They are calculated as

$$F_1D_2 = +0.35 - 0.29 + 0.24 - 0.26 = +0.04,$$

$$F_2D_3 = +0.33 + 0.25 + 0.29 - 0.24 = +0.13,$$

$$F_2S_L = +0 - 0 + 0.29 - 0.24 = +0.05.$$

None of the unused cells has been evaluated favorably and the final solution has been reached. As seen by Table 7-6, 70 units are shipped from F_1 to D_1 and 60 are shipped from F_1 to D_3. From F_2 we shall ship 20 units to D_1 (the Δ is disregarded) and 80 units to D_2. Factory 1 will operate at 86.6% capacity, whereas Factory 2 will operate at full capacity.

We have previously indicated the general nature of the simplex method in Chapter 5. The reader is urged to formulate and solve the illustrative example in this chapter by the simplex method. By this means he will gain proficiency in the simplex technique and, at the same time, will readily recognize the computational efficiency of the transportation method where it is applicable. Note in making this comparison that:

1. All of the exchange coefficients within the simplex table are either 0 or 1. This is a characteristic of linear programming problems that may be placed within the framework of the transportation method.
2. The number of restraints in the simplex table is 5, which is the number of rim requirements minus 1 in the transportation matrix.

THE ASSIGNMENT METHOD

The second special computational technique we shall examine is the assignment method. As was true of the transportation method, the assignment method is computationally more efficient than the simplex method for a special class of linear programming problems. The requirements for using the assignment method can be framed as follows: If there are exactly n jobs and n facilities for accomplishing these jobs, with varying measures of effectiveness associated with each job-facility combination, the assignment method of linear programming will assign the n jobs to the n facilities so as to optimize the given measure of effectiveness. For example, the problem of assigning 4 workers to 4 jobs, where the productivity of each worker on each job is known, can be solved by the assignment method.

The assignment method is best illustrated by means of a simple example. A steamship company has 3 of its ocean-going freighters scheduled to arrive in port on Monday morning. The ships carry cargoes ranging from grain to heavy machinery. There are 4 possible berths at which the vessels can be docked to unload their cargo. Each of the 4 berths is equipped with different cargo-unloading facilities. At some docks the apron or unloading space is very large; at others there is limited heavy equipment. Given the characteristics of the cargoes and the facilities available at each berth the traffic manager has been able to obtain estimates of the number of hours that would be required to unload each of the ships at each of the 4 berths. This information is contained in Table 7-7. The traffic manager must assign the 3 ships to any 3 of the 4 berths in such a manner that the total time required for unloading is minimized.

TABLE 7-7

	Ship		
Berth	1	2	3
A	6	14	20
B	14	11	16
C	12	16	28
D	16	10	7

Before this problem can be solved by the assignment method it must be in a form such that the number of jobs (ships to be unloaded) equals the number of facilities (unloading berths). We can handle this problem by making use of a slack variable. In other words, we shall introduce a dummy ship that has no cargo and therefore does not require any unloading time at any of the 4 berths. Introducing these alterations we obtain the following performance matrix:

	Ship			
Berth	1	2	3	4
A	6	14	20	0
B	14	11	16	0
C	12	16	28	0
D	16	10	7	0

The problem of assigning the ships can easily be solved by inspection, and we can use the results to check our computations by the assignment method. As can be seen from the table, assigning Ship 1 to Berth A, Ship 2 to Berth B, Ship 3 to Berth D, and leaving Berth C unused will minimize total unloading time. The computational technique of the assignment method does not have the intuitive appeal of any of the solution methods so far presented. However, the number of problems that can be solved by this method and its computational efficiency are sufficient to warrant learning the procedure despite its lack of intuitive sophistication.

Computational Procedure

The following is the computational procedure to be followed in using the assignment method.

Step 1. Examine each column of the performance matrix and locate the smallest value in each column. Form a new matrix by subtracting from every value in each column the minimum value of that column. This has been done for the performance matrix, and the results appear in Table 7-8 (the Step 1 reduced matrix). The minimum value used in the subtraction process, which was obtained from the performance matrix, appears in the last row. It is placed there for expository reasons and should not be considered a part of the matrix.

TABLE 7-8

Ship

Berth	1	2	3	4	
A	0	4	13	0	← L_1
B	8	1	9	0	
C	6	6	21	0	
D	10	0	0	0	← L_2
Minimum Value	6	10	7	0	
				↑ L_3	

Step 2. Determine the minimum number of lines that must be drawn through the Step 1 reduced matrix columns and/or rows in order to cover all the zeros in the matrix. In Table 7-8 the arrows to the right and beneath the table indicate that three lines, L_1, L_2, and L_3, drawn through Row 1, Row 4, and Column 4, respectively, will cover all the zeros in the matrix defined by Rows ABCD and Columns 1234.

Step 3. If the number of lines used in Step 2 is equal to the number of facilities (berths), an optimum solution has been reached. The solution could be read from the table by selecting a set of zeros, equal in number to the number of facilities, no two of which appear in the same line. This point may be illustrated by referring to the hypothetical matrix shown in Table 7-9.

TABLE 7-9

	Job				
Facility	A	B	C	D	
E	2	0*	1	4	← L_1
F	0	6	0*	0	← L_2
G	1	0	3	0*	← L_3
H	0*	4	2	0	← L_4

In this illustration the assignments are made in accordance with the procedure outlined above. Job A is assigned to Facility H, Job B to Facility E, Job C to Facility F, and Job D to Facility G. (The assignments are indicated by the asterisks appearing in the respective cells.)

Referring to Table 7-8, we can see that the minimum number of lines needed to cover all zeros is 3, whereas the number of facilities is 4. We have not yet found the optimal solution.

Step 4. Examine the rows of the Step 1 reduced matrix. If all of the rows do not contain at least one zero, proceed in the following manner: select the minimum value in each row and subtract it from every value in the row to form the Step 4 reduced matrix. Using the procedure of covering zeros with lines as described in Step 2, determine whether an optimal solution has been found. If the solution is not optimal, or if there is at least one zero in every row, we proceed to the next step.

Step 5. In our Step 1 reduced matrix there is at least one zero in every row. Therefore, we proceed directly to Step 5 without reducing the matrix as described in Step 4. Examine the Step 1 reduced matrix (or the last reduced matrix in Step 4), and from those values that are not covered by lines choose the minimum value. Since all zeros are covered, the value chosen must be greater than zero. Subtract this value from all of the values in the matrix not covered by lines and add it to all those values that lie at the intersection of the lines. The minimum value not covered by a line is 1 (see Table 7-8). This is subtracted from every value not covered by L_1, L_2, and L_3 and added to every value that lies at the intersection of L_1, L_2, and L_3 (cells A4 and D4). Carrying out this procedure we will obtain the Step 5 reduced matrix shown in Table 7-10.

TABLE 7-10

Ship

Berth	1	2	3	4	
A	0*	4	13	1	← L_1
B	7	0*	8	0	← L_2
C	5	5	20	0*	← L_3
D	10	0	0*	1	← L_4

Referring to the procedure for drawing lines to cover zeros described in Step 2, it can be seen that a minimum of 4 lines is required. According to our criterion for the minimum number of lines being equal to the number of facilities, an optimal solution has been reached. If an optimal solution was not obtained at this stage, Step 5 would be repeated until one was found.

The Solution

As described in Step 3, the solution is determined by selecting a set of zeros from the final matrix (wherever it appears) such that no two zeros chosen will appear in the same line and, further, that the number of zeros chosen be equal to the number of facilities. This has been done in Table 7-10. The choices are shown by the asterisks appearing in the respective cells. The solution arrived at assigns Ship 1 to Berth A, Ship 2 to Berth B, Ship 3 to Berth D; Berth C is not used at all. The total time required to unload all 3 ships is the sum of the times required to unload each ship at its respective berth. Referring to our performance matrix,

$$\text{total time} = 6 + 11 + 7 = 24 \text{ hours.}$$

As in the transportation problem presented earlier, the reader is urged to formulate and solve the ship-scheduling problem by means of the simplex method. This problem may also be solved by the transportation method.

Problems

7-1. Solve the following problem for least cost by the transportation method (costs are per unit).

Shipping From	*Shipping To* 1	2	3	4	*Units Available*
A	0.30	0.20	0.27	0.35	70
B	0.20	0.23	0.32	0.20	20
C	0.15	0.35	0.32	0.40	50
D	0.25	0.27	0.28	0.30	80
E	0.18	0.32	0.28	0.24	60
Requirements	80	60	30	110	280

7-2. The following information is available concerning the operations of the Dorlow Manufacturing Company:

	Month 1	*Month 2*
Units on Order	800	1400
Production Capacity		
Regular time	920	920
Overtime	250	250
Excess cost/unit-overtime	\$1.25	\$1.25
Storage cost/unit	\$0.50	\$0.50

Use the transportation method to solve the problem above.

7-3. Solve Problem 7-2 by the simplex method.

7-4. A drayage company has 4 fully loaded trailers located at points A, B, C, and D. At present the company has 4 tractors located at T_1, T_2, T_3, and T_4. The distances between trailer locations and tractor locations are given by the following matrix:

Trailer	Tractor T_1	T_2	T_3	T_4
A	250	300	175	425
B	190	225	450	300
C	430	320	180	260
D	170	220	400	505

Use the assignment method to determine the assignment of tractors to trailers that minimizes the distance traveled.

7-5. An aircraft manufacturing company has decided to subcontract 3 component parts for its jet aircraft. Three subcontractors have shown an interest in doing the work, although no single firm is able to produce more than one of the orders owing to capacity restrictions. Accordingly, the aircraft manufacturer has asked the 3 firms to submit cost estimates for each of the 3 jobs. These estimates are shown below:

Job	Bidder B_1	B_2	B_3
J_1	175	182	166
J_2	181	180	176
J_3	200	179	172

Use the assignment method to determine which bidder should be given which job. The quality is not expected to differ among the 3 bidders on any job.

7-6. Solve Problem 7-4 by the transportation method.

7-7. Solve Problem 7-5 by the transportation method.

Integer Programming

8

Integer programming is a solution technique applicable to linear programming problems that require integer values for some or all of the variables. Of course, in many problems it is perfectly adequate to round off values because fractional parts are meaningless. For example, had the television assembly problem of Chapter 5 provided the answer that 184.32 black-and-white sets should be produced, we would not worry about dropping the 0.32. But, in other cases, rounding can have serious effects. Suppose the problem were to select the optimum number of airplanes to purchase, and the solution specified 2.39. Rounding down to 2 might be less desirable than some other value. It could even be nonfeasible!

In this chapter we shall illustrate the usefulness of integer programming by applying it to investment planning and then turn to an exposition of Gomory's integer programming solution procedure.

APPLICATION OF INTEGER PROGRAMMING TO INVESTMENT PLANNING

The French have long used linear programming for planning the electric power system in their country.[1] Given that there are a large number of different

[1] See Pierre Massé, *Optimal Investment Decisions* (Englewood Cliffs, N.J.: Prentice-Hall, 1962), p. 160.

methods for generating electricity, and that each method has different performance and cost characteristics, what is the optimal composition of plants of the various types which can satisfy the demand restraints?

This is analogous to the capital budgeting problem of a firm. Martin Weingartner has developed both linear and integer programming models of capital budgeting for the problem of selecting that combination of projects which will maximize the present discounted value of the firm, subject to investment fund limitations.[2]

In this section, we shall construct an example to illustrate the usefulness of integer programming in making investment decisions. The technique for solving integer problems will be deferred until the next section. There, we shall construct a much simpler numerical example and apply the Gomory solution procedure.

Consider the following situation. A firm (or country) is interested in the best combination of electric power plants that will satisfy certain demand and budget restraints. There are two demand restraints:

1. To meet peak demand for electricity during the winter, the power system must have a capacity of at least 3500 megawatts.
2. To satisfy annual demand for electricity, the system must be capable of supplying at least 20 million megawatt-hours.

It should be noted that some types of plants are better suited for meeting restraint 1 (such as seasonal reservoir hydroelectric plants), while others are better suited for meeting restraint 2 (coal burning plants).

There is also a budget restraint. We shall assume that the authorized budget is $400 million. To make the example simple, assume that the types of plants available to the firm are limited to five. These are listed below:

1. 300 megawatt coal burning plant
2. 500 megawatt coal burning plant
3. 1000 megawatt hydroelectric plant
4. 800 megawatt hydroelectric plant
5. 250 megawatt hydroelectric plant

[2] Martin Weingartner, *Mathematical Programming and the Analysis of Capital Budgeting Problems* (Englewood Cliffs, N.J.: Prentice-Hall, 1963).

Some qualifications are needed. The firm can obtain as many of the coal burning plants as it deems desirable, subject to the budget restraint. However, there is only one good location for a large hydroelectric plant. Hence, either type 3 or type 4 may be selected, but not both. Type 5 is a smaller hydroplant which can be employed only if type 4 is selected but not if type 3 is chosen. Of course, it is not necessary to use type 5 even if type 4 is selected.

The following performance and cost characteristics are known:

Plant type	*Annual megawatt-hours (millions)*	*Investment funds required (millions of \$)*	*Total discounted investment and operating cost (millions of \$)*	*Capacity (megawatts)*
1	2.1	33	66	300
2	3.5	50	90	500
3	6.5	150	170	1000
4	6.0	140	160	800
5	1.6	35	40	250

Although the actual numbers are hypothetical, they do indicate the nature of the problem. Some plants conserve on investment funds (coal burning) but have higher operating costs. Also, larger plants tend to be cheaper per unit of capacity, but they are "indivisible." They come only in "lumps," and it may be better to buy a smaller plant to meet a demand deficiency than to buy a larger, lower-unit-cost, plant. And, of course, the plants have different performance characteristics. Thus, it is necessary to use a programming approach, even for this rather simple problem.

The problem can be formulated as follows, where X_i is the number of plants of type i, $(i = 1, 2, \ldots, 5)$:

$$\text{Minimize Cost} = 66X_1 + 90X_2 + 170X_3 + 160X_4 + 40X_5,$$

subject to

$$300X_1 + 500X_2 + 1000X_3 + 800X_4 + 250X_5 \geq 3500, \qquad (1)$$

$$2.1X_1 + 3.5X_2 + 6.5X_3 + 6.0X_4 + 1.6X_5 \geq 20, \qquad (2)$$

$$33X_1 + 50X_2 + 150X_3 + 140X_4 + 35X_5 \leq 400; \qquad (3)$$

$$X_3 + X_4 \leq 1, \qquad (4)$$

$$-X_4 + X_5 \leq 0, \qquad (5)$$

$$X_1, X_2, X_3, X_4, X_5 \geq 0, \qquad (6)$$

$$\text{all } X_i \text{ are integers.} \qquad (7)$$

The final restriction above is that only integer values of the X_i are acceptable. It would clearly be nonsense to determine that 1.72 coal burning plants of 300 megawatts capacity should be ordered. Hence, the problem must be solved by integer programming.

We shall now turn to a brief description of each of the five restraints.

Restraint 1. This is the capacity restraint. The power system must be able to meet a peak demand of 3500 megawatts.

Restraint 2. The power system must also be able to provide 20 million megawatt-hours of electric energy on an annual basis.

Restraint 3. The budget permits a maximum of 400 million dollars to be invested.

Restraint 4. This restraint ensures that the two large hydro plants will not both appear in the optimum solution. Recall that there is only one suitable location for a large hydro plant, and we must explicitly rule out the possibility that both are selected. Since X_3 and X_4 are both nonnegative integers, the restraint permits only three cases:

(a) $X_3 = 0, \quad X_4 = 0$ Neither selected.

(b) $X_3 = 1, \quad X_4 = 0$ Only type 3 selected.

(c) $X_3 = 0, \quad X_4 = 1$ Only type 4 selected.

Restraint 5. Transferring $-X_4$ to the right-hand side, we get $X_5 \leq X_4$. This restraint allows the small hydro plant to be selected only if the large, type 4, hydro plant is selected, as required by the problem specification. Note that the small plant does not need to be chosen even if type 4 is. The three cases permitted are:

(a) $X_4 = 0, \quad X_5 = 0$ Neither selected.

(b) $X_4 = 1, \quad X_5 = 0$ Only the large plant selected.

(c) $X_4 = 1, \quad X_5 = 1$ Both plants selected.

A number of extensions to this problem can be imagined which would make it a more realistic model. Additional types of plants, more demand restraints, and budget restraints for each year are some examples. However, the purpose of the problem is primarily to illustrate the usefulness of integer programming and only secondarily to provide an investment planning model.

OTHER USES OF INTEGER PROGRAMMING

Although integer programming is useful in capital budgeting applications, it is not limited to this field. For example, rather ingenious methods have been devised for using integer programming to solve nonlinear programming problems! For a survey of some of these applications, see G. Hadley, *Nonlinear and Dynamic Programming*.

We now turn to the description of a specific solution procedure for integer programming.

THE GOMORY ALL-INTEGER SOLUTION PROCEDURE

In this section we shall describe the method developed by Ralph E. Gomory in 1958 for solving *all-integer* problems.[3]

There are other algorithms (solution procedures) available for solving *mixed-integer* problems. Since many problems do not require that all of the variables be integers, the serious student will want to examine these. Gomory has developed such a mixed-integer algorithm. A good exposition can be found in Hadley, *Nonlinear and Dynamic Programming*. Another, and considerably different algorithm is of the *branch and bound* type. An introductory discussion of this procedure is available in Hillier and Lieberman, *Introduction to Operations Research*.

Before continuing with the Gomory algorithm we must digress briefly to describe the *dual simplex method*, which is a necessary step in the Gomory method, as we shall see later.

Dual Simplex Method

The dual simplex method is an alternative technique for solving linear programming problems. However, it is appropriate only when the problem is in a

[3] Ralph E. Gomory, "Outline of an Algorithm for Integer Solutions to Linear Programs," *Bulletin of the American Mathematical Society*, LXIV (September 1958), 275–278.

particular form. Suppose we have a maximizing problem and all of the elements in the $C_j - Z_j$ row are zero or negative; yet, we do not have the optimum solution because some of the basic variables are negative (some elements in the P_0 column are negative). This is the situation in which the dual simplex method is applicable.

In essence, the dual simplex method is a procedure which moves from basic solution to basic solution, maintaining only zero or negative elements in the $C_j - Z_j$ row. Ultimately, a basic *feasible* solution is reached (all elements in P_0 column non-negative), and this is also the optimum solution.

The rules for the dual simplex method can be derived by dealing with the primal problem as if the simplex method were being applied simultaneously to the dual problem. We list them below:[4]

> *Rule 1.* Begin with a basic solution with one or more of the elements in the P_0 column negative and all elements in the $C_j - Z_j$ row zero or negative.
>
> *Rule 2.* The leaving variable is the minimum of the negative elements in the P_0 column.
>
> *Rule 3.* Find all negative exchange coefficients in the leaving variable row of the simplex table. Then divide each of these negative coefficients into its corresponding $C_j - Z_j$ element. The variable heading the column with the minimum of these ratios is the entering variable.
>
> *Rule 4.* Having determined the pivot element by rules 2 and 3, proceed with the construction of the next simplex table exactly as in the simplex method.

We now turn to an example which illustrates the Gomory all-integer algorithm. The dual simplex method rules will be applied in this example.

Application of Gomory Algorithm

First, we shall describe the logic of the procedure. Take as the starting point the optimum solution to the linear programming problem, where the integrality requirements are ignored. Now, if the solution is, by chance, an integer solution, the

[4] See G. Hadley, *Linear Programming* (Reading, Mass.: Addison-Wesley Publishing Co., 1962), p. 242, for a derivation of the rules.

integer problem is solved. Usually, however, this is not the case, and an *additional* linear constraint is incorporated into the problem. This additional constraint, called a "cutting-plane," cuts away the noninteger optimum solution from the feasible region. But, it does not cut away any feasible *integer* solutions. The new constraint is such that the dual simplex method becomes appropriate. Hence, the linear programming problem is reoptimized via the dual simplex method. If the new optimum solution is all-integer, the solution has been reached; if not, another cutting plane is constructed and the procedure repeated. This process continues until the optimum integer solution is reached.

As a simple example, find the *integer values* of X and Y which maximize

$$Z = 3X + 4Y$$

subject to

$$1X + 2Y \leq 8,$$
$$4X + 1Y \leq 16,$$

and

$$X,\ Y \geq 0.$$

Introducing slack variables, S_1 and S_2, and solving by the simplex method give the optimum solution shown in Table 8-1.

Ignoring the integrality requirements, the optimum solution is

$$X = 3\frac{3}{7}, \quad Y = 2\frac{2}{7} \quad \text{and} \quad Z = 3\left(3\frac{3}{7}\right) + 4\left(2\frac{2}{7}\right) = 19.43.$$

TABLE 8-1

		Column	1	2	3	4	
		C_j	3	4	0	0	
Row	C_i	*Solution*	X	Y	S_1	S_2	P_0
1	4	Y	0	1	$8/14$	$-1/7$	$2\,2/7$
2	3	X	1	0	$-1/7$	$2/7$	$3\,3/7$
		Z_j	3	4	$15/7$	$2/7$	19.43
		$C_j - Z_j$	0	0	$-15/7$	$-2/7$	

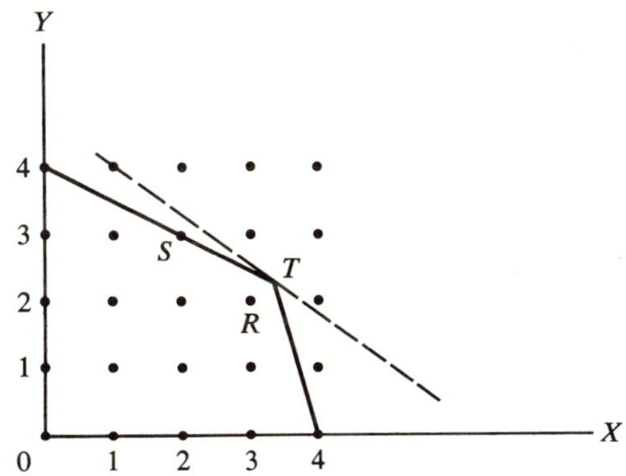

Figure 8-1. Graphical Solution.

The graphical solution is shown as point T in Figure 8-1. The dashed line is the Z function. The black dots indicate integer solutions.

Consider what would happen if we simply rounded off the solution to meet the integrality requirement. The solution would then become

$$X = 3, \quad Y = 2 \quad \text{and} \quad Z = 3(3) + 4(2) = 17.$$

In this case, the rounded solution is feasible (point R in Figure 8-1), but this is not always the case. Even so, the rounded solution is *not* the optimal integer solution. If the dashed Z function moved leftward parallel to itself, it would reach point S before it would reach point R. Hence, the optimal integer solution is point S:

$$X = 2, \quad Y = 3 \quad \text{and} \quad Z = 3(2) + 4(3) = 18.$$

Constructing the Cutting-Plane

The next step is to construct a new linear restraint which will cut away the optimal noninteger solution from the feasible region. The restraint must not cut away any feasible integer solutions. We shall show the graphical version of the cutting plane first and then show how the algebra is accomplished.

Figure 8-2 shows that the new restraint, CP_1 (cutting plane) reduces the feasible region by a tiny triangular area which includes the point T.

To derive the restraint algebraically, return to Table 8-1 and select the variable

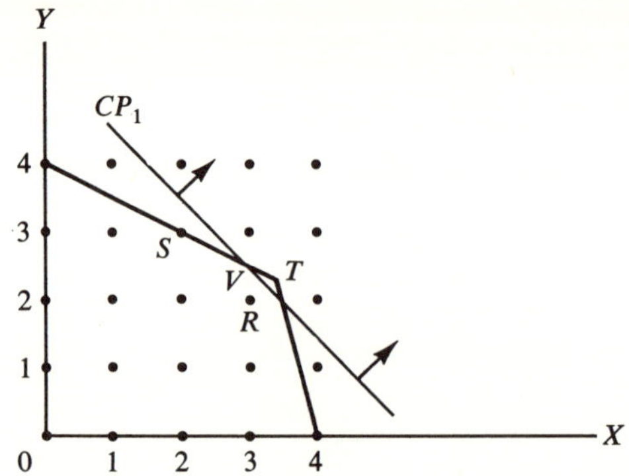

Figure 8-2.

with the largest fractional part, or X. Write the equation of that row of the table:

$$1X + 0Y - \frac{1}{7}S_1 + \frac{2}{7}S_2 = 3\frac{3}{7}. \tag{8}$$

Next, break up all coefficients and the constant into an integer and a non-negative fractional part:

$$1X + 0Y + \left(-1 + \frac{6}{7}\right)S_1 + \left(0 + \frac{2}{7}\right)S_2 = \left(3 + \frac{3}{7}\right).$$

We can rearrange the equation as follows:

$$X = \frac{3}{7} - \frac{6}{7}S_1 - \frac{2}{7}S_2 + 3 + 1S_1. \tag{9}$$

We have now collected all the terms involving fractions as the first three terms on the right-hand side, and the two integer terms are placed together on the far right.

The major idea in the derivation is that, for X to be an integer, we need consider only the three fractional terms; the terms on the far right will be integers for any integer solution just as they are. Thus, for X to be an integer, the sum of the first three terms,

$$\frac{3}{7} - \frac{6}{7}S_1 - \frac{2}{7}S_2 \tag{10}$$

must also be an integer. Using this information, we can infer that

$$\frac{3}{7} - \frac{6}{7} S_1 - \frac{2}{7} S_2 \leq 0. \tag{11}$$

The *only* way (10) can be positive for integer values of S_1 and S_2 is as a *fraction*. For example, if $S_1 = 0$ and $S_2 = 1$, then (10) is $3/7 - 2/7 = 1/7$. And, if (10) is a fraction, then X is noninteger—a possibility which we wish to guard against.

As a consequence of breaking up the coefficients and constant into an integer and a non-negative fractional part, the cutting-plane restraint will always be of the form illustrated by (11). That is, the form will be a positive fraction minus a number of terms with fractional coefficients, and which is restrained to be less than or equal to zero.

Adding a slack variable, S_3, to (11), we can write the new restraint as an equation:

$$-\frac{6}{7} S_1 - \frac{2}{7} S_2 + 1S_3 = -\frac{3}{7}. \tag{12}$$

Equation (12) must be converted into the variables X and Y in order to graph the equation in Figure 8-2. Substitute $S_1 = 8 - X - 2Y$ and $S_2 = 16 - 4X - Y$ into (12) and the result is equation (13):

$$-\frac{6}{7}(8 - X - 2Y) - \frac{2}{7}(16 - 4X - Y) = -\frac{3}{7} \tag{13}$$

We set $S_3 = 0$ in order to graph the restraint line (as in Chapter 4). Combining terms gives

$$X + Y = 5\tfrac{1}{2}$$

which is graphed in Figure 8-2 as the cutting plane, CP_1.

Finally, we append equation (12) to the other two equations given in Table 8-1. The variable S_3 becomes a basic variable, and its value is given in the P_0 column. The new simplex table is shown as Table 8-2.

Notice that all elements in the $C_j - Z_j$ row are either zero or negative. Yet, the solution is not feasible because $S_3 = -3/7$, a negative value. Hence, the problem is now in the appropriate form to apply the dual simplex method.

Rule 2 of the dual simplex method, stated earlier in this chapter, indicates that the variable S_3 should be the leaving variable (the minimum of negative elements in column P_0).

TABLE 8-2

		Column	1	2	3	4	5	
		C_j	3	4	0	0	0	
Row	C_i	Solution	X	Y	S_1	S_2	S_3	P_0
1	4	Y	0	1	$8/14$	$-1/7$	0	$16/7$
2	3	X	1	0	$-1/7$	$2/7$	0	$24/7$
3	0	S_3	0	0	$-6/7$	$-2/7$	1	$-3/7$
		Z_j	3	4	$15/7$	$2/7$	0	19.43
		$C_j - Z_j$	0	0	$-15/7$	$-2/7$	0	

Rule 3 directs us to select the coefficients in the columns headed by variables S_1 and S_2 and in the S_3 row (because they are negative). Dividing these coefficients into their respective $C_j - Z_j$ elements, we obtain for the S_1 column

$$\frac{-15/7}{-6/7} = \frac{15}{6},$$

and for the S_2 column

$$\frac{-2/7}{-2/7} = 1.$$

Since the ratio associated with the S_2 column is the minimum, S_2 is selected as the entering variable. The pivot element is the $-2/7$ in the entering variable column and leaving variable row. By constructing the new simplex table, exactly as in the simplex method, we obtain Table 8-3.

Again, we have an optimal solution:

$$X = 3,$$

$$Y = \frac{5}{2},$$

$$S_2 = \frac{3}{2},$$

and

$$Z = 3(3) + 4\left(\frac{5}{2}\right) = 19.$$

TABLE 8-3

Row	C_i	Solution	X	Y	S_1	S_2	S_3	P_0
		Column	1	2	3	4	5	
		C_j	3	4	0	0	0	
1	4	Y	0	1	1	0	$-\frac{1}{2}$	$\frac{5}{2}$
2	3	X	1	0	-1	0	1	3
3	0	S_2	0	0	3	1	$-\frac{7}{2}$	$\frac{3}{2}$
		Z_j	3	4	1	0	1	19
		$C_j - Z_j$	0	0	-1	0	-1	

This corresponds to point V in Figure 8-2. However, the solution is not an integer solution. We must repeat the procedure—construct a new cutting plane and re-optimize by the dual simplex method.

Before proceeding with the algebraic method, we again show the preliminary graphical version in Figure 8-3. Both cutting planes are shown. The latest restraint, labeled CP_2, passes through points S and R. The new cutting plane cuts away the previous optimum point V but does not cut away any feasible integer points.

Algebraically, we select the variable in Table 8-3 with the largest fractional part and write its equation. Since there is a tie for the largest fractional part, we shall

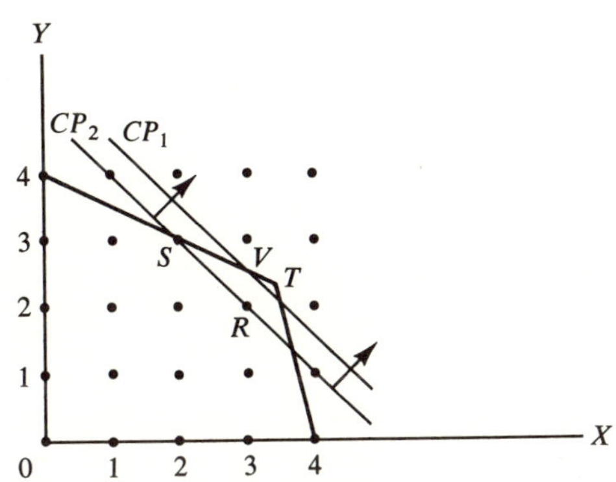

Figure 8-3.

arbitrarily choose S_2:

$$3S_1 + 1S_2 - \frac{7}{2}S_3 = \frac{3}{2}. \tag{14}$$

Breaking up the coefficients and the constant into an integer and a non-negative fractional part, we get

$$3S_1 + 1S_2 + \left(-4 + \frac{1}{2}\right)S_3 = \left(1 + \frac{1}{2}\right).$$

Rearranging the equation we get

$$S_2 = \frac{1}{2} - \frac{1}{2}S_3 + 1 - 3S_1 + 4S_3.$$

Hence

$$\frac{1}{2} - \frac{1}{2}S_3 \leq 0 \tag{15}$$

is the new restraint. Adding a slack variable, S_4, we get equation (16):

$$-\frac{1}{2}S_3 + S_4 = -\frac{1}{2}. \tag{16}$$

Equation (16) can be expressed in terms of X and Y as

$$X + Y = 5.$$

This is the equation graphed as CP_2 in Figure 8-3. Appending equation (16) to the other three equations given in Table 8-3, we can construct Table 8-4.

TABLE 8-4

		Column	1	2	3	4	5	6	
		C_j	3	4	0	0	0	0	
Row	C_i	*Solution*	X	Y	S_1	S_2	S_3	S_4	P_0
1	4	Y	0	1	1	0	$-1/2$	0	$5/2$
2	3	X	1	0	-1	0	1	0	3
3	0	S_2	0	0	3	1	$-7/2$	0	$3/2$
4	0	S_4	0	0	0	0	$-1/2$	1	$-1/2$
		Z_j	3	4	1	0	1	0	19
		$C_j - Z_j$	0	0	-1	0	-1	0	

When we apply the rules of the dual simplex method, we find that S_3 is the entering variable and S_4 is the leaving variable. The pivot element is, therefore, $-\frac{1}{2}$. Applying the simplex routine to construct the new simplex table yields Table 8-5.

TABLE 8-5

		Column	1	2	3	4	5	6	
		C_j	3	4	0	0	0	0	
Row	C_i	*Solution*	X	Y	S_1	S_2	S_3	S_4	P_0
1	4	Y	0	1	1	0	0	−1	3
2	3	X	1	0	−1	0	0	2	2
3	0	S_2	0	0	3	1	0	−7	5
4	0	S_3	0	0	0	0	1	−2	1
		Z_j	3	4	1	0	0	2	18
		$C_j - Z_j$	0	0	−1	0	0	−2	

We have finally reached the optimum integer solution. All elements in the $C_j - Z_j$ row are zero or negative, and all values in the P_0 column are positive integers. The solution is

$$X = 2$$

$$Y = 3$$

$$S_2 = 5$$

$$S_3 = 1$$

and

$$Z = 3(2) + 4(3) = 18.$$

This corresponds to point S in Figure 8-3.

One concluding observation is that this algorithm is rather inefficient in comparison with the simplex method. For the purposes of our extremely simple example, we needed a relatively large number of simplex tables. In any case, the need for more efficient integer programming algorithms is apparent, and considerable research is currently being directed toward this end.

Problems

8-1. Use the Gomory method to find the integer values of X and Y which will maximize $Z = 3X + 4Y$, subject to

$$2X + 3Y \leq 9,$$

$$2X + Y \leq 6,$$

$$X,\ Y \geq 0.$$

8-2. Solve the electric power investment planning problem formulated in the chapter.

8-3. The cost function of a plant is

$$C = \alpha + \beta X \quad \text{if } X > 0,$$

$$C = 0 \qquad\qquad \text{if } X = 0,$$

where C is total cost and X is the rate of output (α and β are positive constants). Thus, if the plant is producing at a positive output rate, there is a fixed cost, α, which is independent of the level of output. The fixed cost can be avoided only if the plant is shut down.

The cost function is nonlinear as it stands and cannot be the objective function of a *linear* programming minimizing problem. Introduce an integer variable, I, which can take on the values of 0 and 1 only, and transform the cost function into a linear one. *Hint:* Let α be the coefficient of I. Construct restraints which require I to equal 1 if $X > 0$. Why is it not necessary to restrict I to equal zero if $X = 0$ (in a cost minimizing linear programming problem)?

9 Input-Output Analysis

Input-output analysis is primarily due to the work of W. W. Leontief.[1] The basic idea underlying input-output analysis is that there is a high degree of production interdependence among the goods and services produced in an industrial economy. Therefore, an increase in the output of television sets for consumer final demand implies an increase in the output of many other goods needed in the production of television sets. Furthermore, it may be true that this induced increase in the output of other goods may, in turn, increase further the output of television sets. In essence, Leontief proposed that the production of all industries in an economy should be treated as a simultaneous system of linear equations.

Our discussion will be concerned with the use of input-output analysis as a forecasting tool. There are other uses for input-output analysis in aggregative economic planning, but we shall not consider these here. The interested reader might wish to examine H. B. Chenery and P. G. Clark, *Interindustry Economics*.

Input-output analysis is useful at many levels of aggregation. A firm in a particular industry might be concerned only with industries that are closely related

[1] W. W. Leontief, "Quantitative Input and Output Relations in the Economic System of the United States," *Review of Economics and Statistics*, August 1936.

to its own, whereas an economy-wide planning agency might be interested in all producing sectors in the economy. For example, an input-output table consisting of eighty-seven producing sectors has been constructed for the United States.[2] A firm, on the other hand, might wish to choose *one* of the eighty-seven sectors and disaggregate that particular sector into more meaningful subsectors, or industries.

The plan for this chapter is to illustrate the basic concepts of input-output analysis by a simple three-industry example. Larger models are obviously more realistic, but the complexity of such calculations is too great for expositional purposes. Before turning to the example, however, we must digress to provide an introduction to matrix algebra.

MATRIX ALGEBRA

This section is a logical continuation of Chapter 4 in which we discussed the Gauss-Jordan method for solving simultaneous linear equations. Here we shall extend that discussion to include matrices. The matrix discussion will be limited to basic concepts necessary for the input-output analysis. An excellent book on matrices, designed for students of management and economics, is G. Hadley, *Linear Algebra.*

To illustrate the concepts we will use the two-equation two-variable example solved in Chapter 4:

$$4X_1 + 8X_2 = 10, \tag{1}$$

$$1X_1 - 0.5X_2 = 1. \tag{2}$$

You will recall our earlier reference to the *coefficient matrix* and the *identity matrix*. The coefficient matrix of equations (1) and (2) is

$$\begin{bmatrix} 4 & 8 \\ 1 & -0.5 \end{bmatrix}.$$

Thus, a matrix is simply a rectangular array of numbers. In general, a matrix can have any number of rows and columns and it need not be square. For example, an $m \times n$ matrix is a matrix with m rows and n columns. Let **A** be a general $m \times n$

[2] See the *Survey of Current Business*, U.S. Department of Commerce, September 1965.

matrix (capital letters will always be used to denote matrices):

$$\mathbf{A} = \begin{bmatrix} a_{11} & a_{12} & \cdots & a_{1n} \\ a_{21} & a_{22} & \cdots & a_{2n} \\ \vdots & \vdots & & \vdots \\ a_{m1} & a_{m2} & \cdots & a_{mn} \end{bmatrix},$$

The double subscript denotes the location of the element. The element a_{ij} is located in the ith row and jth column.

Matrix Operations

We now wish to illustrate the matrix operations of addition and multiplication. *Matrix addition* is possible only if the two matrices have the same *dimension*—that is, the same number of rows and columns. Two matrices which have the same dimension are added by the addition of corresponding elements. For example,

$$\begin{bmatrix} 1 & 2 \\ 3 & 1 \end{bmatrix} + \begin{bmatrix} 0 & -2 \\ 4 & 3 \end{bmatrix} = \begin{bmatrix} 1 & 0 \\ 7 & 4 \end{bmatrix},$$

Matrix multiplication is possible only when the two matrices are *conformable*. The conformability condition is that the product of two matrices, **AB** is obtainable only if the number of columns of **A** equals the number of rows of **B**. Otherwise, **AB** is not defined. If **A** is $m \times p$ and **B** is $p \times n$, then the product **AB** is defined. The product matrix **AB** will be of dimension $m \times n$ with any element, c_{ij}, determined as follows:

$$c_{ij} = a_{i1}b_{1j} + a_{i2}b_{2j} + \cdots + a_{ip}b_{pj}.$$

Example.

$$\mathbf{C} = \begin{bmatrix} 2 & 1 & 1 \\ -1 & 2 & 3 \end{bmatrix}, \qquad \mathbf{D} = \begin{bmatrix} 0 & 4 \\ -2 & 1 \end{bmatrix}.$$

CD is not defined

$$\mathbf{DC} = \begin{bmatrix} 0 & 4 \\ -2 & 1 \end{bmatrix} \begin{bmatrix} 2 & 1 & 1 \\ -1 & 2 & 3 \end{bmatrix} = \begin{bmatrix} a & c & e \\ b & d & f \end{bmatrix},$$

where

$$\begin{aligned} a &= 0(2) + 4(-1) = -4, \\ b &= -2(2) + 1(-1) = -5, \\ c &= 0(1) + 4(2) = 8, \\ d &= -2(1) + 1(2) = 0, \\ e &= 0(1) + 4(3) = 12, \\ f &= -2(1) + 1(3) = 1, \end{aligned}$$

or

$$\mathbf{DC} = \begin{bmatrix} -4 & 8 & 12 \\ -5 & 0 & 1 \end{bmatrix}.$$

Recall that the identity matrix, discussed in Chapter 4, was a square matrix of the form

$$\mathbf{I} = \begin{bmatrix} 1 & 0 & 0 \\ 0 & 1 & 0 \\ 0 & 0 & 1 \end{bmatrix}.$$

The above identity matrix is of the third order, since it has three rows and three columns. As a further illustration of matrix multiplication, we shall multiply the coefficient matrix, **A**, by a conformable identity matrix (a second-order identity matrix):

$$\mathbf{AI} = \begin{bmatrix} 4 & 8 \\ 1 & -0.5 \end{bmatrix} \begin{bmatrix} 1 & 0 \\ 0 & 1 \end{bmatrix} = \begin{bmatrix} 4 & 8 \\ 1 & -0.5 \end{bmatrix}.$$

Observe that $\mathbf{AI} = \mathbf{A}$ and, inversely, that $\mathbf{IA} = \mathbf{A}$.

Matrix Inverse

The *inverse* of the matrix **A** is written as $\mathbf{A}^{-1}$. The inverse matrix is unique and has the following property:

$$\mathbf{AA}^{-1} = \mathbf{A}^{-1}\mathbf{A} = \mathbf{I}.$$

For example, consider the matrix **C**:

$$\mathbf{C} = \begin{bmatrix} 0.05 & 0.8 \\ 0.1 & -0.4 \end{bmatrix}.$$

Since

$$\mathbf{AC} = \begin{bmatrix} 4 & 8 \\ 1 & -0.5 \end{bmatrix} \begin{bmatrix} 0.05 & 0.8 \\ 0.1 & -0.4 \end{bmatrix} = \begin{bmatrix} 1 & 0 \\ 0 & 1 \end{bmatrix} = \mathbf{I},$$

and

$$\mathbf{CA} = \begin{bmatrix} 0.05 & 0.8 \\ 0.1 & -0.4 \end{bmatrix} \begin{bmatrix} 4 & 8 \\ 1 & -0.5 \end{bmatrix} = \begin{bmatrix} 1 & 0 \\ 0 & 1 \end{bmatrix} = \mathbf{I},$$

then $\mathbf{C}$ is the inverse matrix of $\mathbf{A}$:

$$\mathbf{C} = \mathbf{A}^{-1}.$$

Solution of Matrix Equation

Not all matrices have inverses. In fact, only certain *square* matrices have inverses. To understand the properties of those matrices that do have inverses (called *nonsingular* matrices), we shall return to our example, equations (1) and (2).

$$4X_1 + 8X_2 = 10, \tag{1}$$

$$1X_1 - 0.5X_2 = 1. \tag{2}$$

Written in matrix notation, we have

$$\mathbf{Ax} = \mathbf{b}, \tag{3}$$

where

$$\mathbf{x} = \begin{bmatrix} X_1 \\ X_2 \end{bmatrix} \quad \text{and} \quad \mathbf{b} = \begin{bmatrix} 10 \\ 1 \end{bmatrix}.$$

$\mathbf{A}$ is the coefficient matrix. Matrices, such as $\mathbf{x}$ and $\mathbf{b}$, which have only one column (or one row) are usually called *vectors*. (Lowercase letters are used to represent vectors.) Vectors can be interpreted as matrices insofar as the operations of addition and multiplication are concerned.

Based on our earlier discussion, we can insert an identity matrix in equation (3) without altering the equation:

$$\mathbf{Ax} = \mathbf{Ib}. \tag{4}$$

To solve for the vector $\mathbf{x}$, we multiply both sides of (4) by $\mathbf{A}^{-1}$:

$$\mathbf{A}^{-1}\mathbf{Ax} = \mathbf{A}^{-1}\mathbf{Ib}.$$

Since $\mathbf{A}^{-1}\mathbf{A} = \mathbf{I}$,

$$\mathbf{I}\mathbf{x} = \mathbf{A}^{-1}\mathbf{I}\mathbf{b} \tag{5}$$

and, dropping the identity matrices, we have

$$\mathbf{x} = \mathbf{A}^{-1}\mathbf{b}. \tag{6}$$

Since $\mathbf{A}^{-1}$ is known, we can write (6) as

$$\begin{bmatrix} X_1 \\ X_2 \end{bmatrix} = \begin{bmatrix} 0.05 & 0.8 \\ 0.1 & -0.4 \end{bmatrix} \begin{bmatrix} 10 \\ 1 \end{bmatrix} = \begin{bmatrix} 1.3 \\ 0.6 \end{bmatrix}.$$

Thus, the solution is

$$X_1 = 1.3; \qquad X_2 = 0.6.$$

Two points should be made. Equation (6) shows that matrix notation provides a concise way of writing the solution to a system of equations. Equation (6) also indicates that if the inverse matrix is known, it is easy to obtain the new solution when $\mathbf{b}$ is changed. (This is useful in input-output analysis.)

Finding the Inverse

You may wonder why we added an $\mathbf{I}$ in (4) and then dropped it in (6). The reason was simply to help with the exposition in this section, where we shall demonstrate one method for finding the inverse matrix. The method is to apply the Gauss-Jordan procedure to equation (4). (You may wish to review the Gauss-Jordan discussion in Chapter 4 before continuing.)

Note carefully what happened to equation (4) when we multiplied both sides by $\mathbf{A}^{-1}$. First, consider the left-hand side. The coefficient matrix, $\mathbf{A}$, was transformed into the identity matrix, as shown in equation (5). But this is exactly what the Gauss-Jordan procedure does in finding the solution to a system of equations. Hence, we conclude that multiplying the coefficient matrix by $\mathbf{A}^{-1}$ is equivalent to applying the rather tedious operations of the Gauss-Jordan routine.

Now consider what happens on the right-hand side of (4) when it is multiplied by $\mathbf{A}^{-1}$. As shown in equation (6), the identity matrix is transformed into $\mathbf{A}^{-1}$. Since we have observed that the Gauss-Jordan routine is equivalent to multiplying by $\mathbf{A}^{-1}$, we are now able to devise a method for finding $\mathbf{A}^{-1}$; we construct a table with the

A matrix on the left and the **I** matrix on the right (corresponding to equation (4)):

$$\begin{bmatrix} 4 & 8 \\ 1 & -0.5 \end{bmatrix} \qquad \begin{bmatrix} 1 & 0 \\ 0 & 1 \end{bmatrix}.$$

Next, apply the Gauss-Jordan routine to **A**, thereby transforming it into the identity matrix. Apply the same operations simultaneously to the identity matrix on the right. Then, when **A** becomes **I** on the left. **I** will become $\mathbf{A}^{-1}$ on the right!

Divide the first row of both matrices by 4. The result is

$$\begin{bmatrix} 1 & 2 \\ & \end{bmatrix} \qquad \begin{bmatrix} 0.25 & 0 \\ & \end{bmatrix}.$$

Multiply the new first rows above by -1 and add the results to the second rows.

$$\begin{bmatrix} 1 & 2 \\ 0 & -2.5 \end{bmatrix} \qquad \begin{bmatrix} 0.25 & 0 \\ -0.25 & 1 \end{bmatrix}.$$

To complete the transformation of the matrix on the left into the identity matrix, the second row must be divided by -2.5. Then, we must multiply the resulting second row by -2 and add that result to the first row. By applying the same operations to the matrix on the right, we obtain

$$\begin{bmatrix} 1 & 0 \\ 0 & 1 \end{bmatrix} \qquad \begin{bmatrix} 0.05 & 0.8 \\ 0.1 & -0.4 \end{bmatrix}.$$

The matrix on the right is the inverse of **A**, or

$$\mathbf{A}^{-1} = \begin{bmatrix} 0.05 & 0.8 \\ 0.1 & -0.4 \end{bmatrix}.$$

We can now understand which square matrices have inverses—only those that can be transformed by Gauss-Jordan operations into identity matrices; they are known as *nonsingular* matrices. The coefficient matrix of the example in Chapter 4 which demonstrated an infinite number of solutions is

$$\begin{bmatrix} 4 & 8 \\ 2 & 4 \end{bmatrix}.$$

Recall that this matrix cannot be transformed into an identity matrix because one row is a multiple of the other; that is, the rows are not independent. The attempt to do so led to a matrix with all zeros in the second row:

$$\begin{bmatrix} 1 & 2 \\ 0 & 0 \end{bmatrix}.$$

Hence, it does not have an inverse and is called a *singular* matrix.

One of the important results reached in Chapter 4 can now be restated in matrix language. If the coefficient matrix $\mathbf{A}$ of a matrix equation, $\mathbf{Ax} = \mathbf{b}$, is nonsingular, a unique solution vector, $\mathbf{x}$, exists. However, if $\mathbf{A}$ is singular, an infinite number of solution vectors, $\mathbf{x}$, are possible.

Determinants

Although we do not believe that determinants are important enough for our discussion of input-output analysis to warrant a detailed exposition, we shall state the determinantal test for nonsingularity.

The determinant of the matrix $\mathbf{A}$, denoted $|\mathbf{A}|$, is shown below for two cases:

1. *If* $\mathbf{A}$ *is a* 2×2 *matrix*, then

$$|\mathbf{A}| = a_{11}a_{22} - a_{21}a_{12}.$$

2. *If* $\mathbf{A}$ *is a* 3×3 *matrix*, then

$$|\mathbf{A}| = a_{11}a_{22}a_{33} + a_{12}a_{23}a_{31} + a_{21}a_{32}a_{13} - a_{31}a_{22}a_{13} - a_{21}a_{12}a_{33} - a_{32}a_{23}a_{11}.$$

Test: If the determinant of a matrix is zero, the matrix is singular; if the determinant s nonzero, the matrix is nonsingular. Thus, the matrix

$$\begin{bmatrix} 4 & 8 \\ 1 & -0.5 \end{bmatrix}$$

is nonsingular, because

$$\begin{vmatrix} 4 & 8 \\ 1 & -0.5 \end{vmatrix} = (4)(-0.5) - (1)(8) = -10.$$

And the matrix

$$\begin{bmatrix} 4 & 8 \\ 2 & 4 \end{bmatrix}$$

is singular, because

$$\begin{vmatrix} 4 & 8 \\ 2 & 4 \end{vmatrix} = (4)(4) - (2)(8) = 0.$$

Summary

If we have a linear equation system

$$\mathbf{Ax} = \mathbf{b}$$

and $\mathbf{A}$ is nonsingular (implying $|\mathbf{A}| \neq 0$), then $\mathbf{A}^{-1}$ exists, and the unique solution is

$$\mathbf{x} = \mathbf{A}^{-1}\mathbf{b}.$$

INPUT-OUTPUT MODEL

For simplicity, consider an economy that consists solely of three industries. Each industry produces a single, homogeneous commodity. The transactions in this economy can be described by an input-output table (Table 9-1).

TABLE 9-1. Transactions Table.

From	*To* *A*	*B*	*C*	*Consumers*	*Total*
A	40	10	30	60	140
B	10	20	10	80	120
C	20	10	10	40	80
Households	70	80	30	0	180
Total	140	120	80	180	

The entries in the table are all in terms of dollars. Thus, the first row (row *A*) indicates that industry *A* produced \$140 worth of goods in the period covered by the table. Of the \$140 worth of goods, industries *A*, *B*, and *C* used \$40, \$10, and \$30 worth of commodity *A*, respectively, for inputs into their own production processes.[3] Consumers bought the remaining \$60 worth of *A* for their consumption.

A simple but basic point is that to produce \$60 worth of output *A* for consumer final demand, industry *A* must produce a total output of \$140. The reason is that \$80 worth of *A* serves as raw materials.

Rows *B* and *C* can be interpreted similarly. The Households row is a "catch-all" which includes wages, salaries, property income, and so forth. Households can be considered to be a fourth industry which produces labor and managerial and capital services. Industries *A*, *B*, and *C* require \$70, \$80, and \$30 worth of Households' output, respectively, for their production processes.

Now that we have looked at the rows of Table 9-1, representing the allocation of total output of each industry, we turn to the meaning of the columns. Column *A* has the same total, \$140, as row *A*. This is perfectly reasonable because we wish to think of the columns as the total input distribution. Thus, industry *A* purchases \$40 worth of output from industry *A*, \$10 worth of output *B*, and \$20 worth of output *C*, for use in its own production process. And, as mentioned above, industry *A* requires \$70 worth of labor, managerial, and capital services. Hence, total input of \$140 (the column sum) must equal total output of \$140 (the row sum).

Columns *B* and *C* can be interpreted similarly. The Consumers column is sometimes called the "final bill of goods." It represents the final consumption of each output by consumers. The total is, of course, equal to the total payments to Households, or \$180.

Forecasting Problem

Having described the three-industry economy, we can now turn to input-output *analysis*. First, we need to pose the forecasting question that we wish to answer.

Assume that a firm in industry *B* wishes to forecast what its production requirements will be three years from now. The firm must make an estimate of its

[3] The fact that industry *A* uses \$40 worth of its own output in its production process should not be difficult to understand. An electric power company, for example, needs electricity to operate its own plants, office buildings, etc.

share of the market based upon marketing information. However, input-output analysis can be used to forecast the *total* demand for commodity A, which is the task before us.

The final bill of goods for the current year can be written as a *final demand vector* (the Consumers column in Table 9-1):

$$\mathbf{b}_0 = \begin{bmatrix} 60 \\ 80 \\ 40 \end{bmatrix}.$$

Marketing forecasts for the elements of the final demand vector three years hence are required. For example, the trend in consumer demand for A may be down, while consumer demand for B increases rapidly, and demand for C increases, but less rapidly. These considerations might lead to a final demand vector forecast of, say:

$$\mathbf{b}_3 = \begin{bmatrix} 50 \\ 100 \\ 45 \end{bmatrix}.$$

Given that total output of B is \$120 now, what will it be in three years? The essence of our method of solution is that we shall explicitly take into account the fact that industry B is a part of an interrelated economy and that changes in final demand for commodities A and C indirectly affect the demand for total output of B. In brief, the total demand for B in three years depends not only on consumer demand for B, but also on consumer demand for A and C.

Assumptions

In order to use the input-output method for forecasting, we must make several assumptions:

1. As stated at the outset, each industry is assumed to produce a single homogeneous commodity. (It is possible to define composite commodities. For example, a composite commodity might be two automobiles and one truck.)

2. Production in each industry is characterized by fixed proportions and constant returns to scale. To illustrate, let prices be given and interpret "a dollar's worth" of each commodity as its unit. To produce 140 units of A, the production process requires 40, 10, and 20 units of A, B, and C, respectively. The assumption of fixed proportions and constant returns to scale means that each unit of A requires

$$\frac{40}{140} = 0.286 \text{ units of } A,$$

$$\frac{10}{140} = 0.071 \text{ units of } B,$$

$$\frac{20}{140} = 0.143 \text{ units of } C.$$

Thus, for an output of A of, say X_1, the total input of A required is $0.286\ X_1$; the total input of B required is $0.071\ X_1$; and of C is $0.143\ X_1$.

3. Technological change is assumed to be absent. Thus, the technical coefficients in assumption 2 do not change over the period of analysis.

Technical Coefficients

We can completely describe the production processes of the three industry economy by a table of technical coefficients. The technical coefficients were just calculated for industry A and appear in Column A of Table 9-2.

Columns B and C can be calculated similarly. Thus, divide each entry in Column B of Table 9-1 by the total output of industry B, and do the same for Column C.

TABLE 9-2. Technical Coefficients Table.

	A	B	C
A	0.286	0.083	0.375
B	0.071	0.167	0.125
C	0.143	0.083	0.125

The reader should note the similarity between the technical coefficients above and the exchange coefficients of linear programming. In fact, input-output analysis can be considered a special case of linear programming.[4] But this should become more clear as we proceed.

Forecast of Total Outputs

Marketing studies show us that the final demand vector forecast for three years hence will be

$$\mathbf{b}_3 = \begin{bmatrix} 50 \\ 100 \\ 45 \end{bmatrix}.$$

What will be the total output of industry B in three years if $\mathbf{b}_3$ is correct? To answer this question, we must solve a system of simultaneous linear equations.

Let the unknown total outputs of industries A, B, and C be represented by X_1, X_2, and X_3, respectively. Thus, for the *current* year, Table 9-1 shows that:

$$X_1 = 140,$$
$$X_2 = 120,$$
$$X_3 = 80.$$

We shall also need to refer to the *total output vector:*

$$\mathbf{x} = \begin{bmatrix} X_1 \\ X_2 \\ X_3 \end{bmatrix}.$$

The problem is to find the vector $\mathbf{x}$ for three years hence.

[4] Specifically, input-output analysis is a linear programming problem in which there is no choice. There are n variables in n equations and therefore the feasible region is a single point—the unique solution to the system. An objective function would be irrelevant, since there is no choice to be made among feasible solutions. Input-output models are sometimes converted into linear programming models by introducing alternative production processes and by introducing land, labor, and capital-resource restraints. This is usually done in economic planning models for developing countries. For a simple example, see the Maluland case at the end of this book. For more advanced reading, see Chenery and Clark, *Interindustry Economics*.

We can now write the first equation. Consulting Table 9-2, we allocate the output of industry A as follows:

$$0.286X_1 + 0.083X_2 + 0.375X_3 + 50 = X_1. \qquad (1)$$

The first three terms represent the amounts of commodity A required by industries A, B, and C, respectively. The constant, 50, is the forecast final demand for A. Hence, the sum of the input requirements for A going to industries A, B, and C, and of the consumer demand for A, must equal the total output of A, or X_1.

Two other equations can be written to describe the allocation of total output of industries B and C:

$$0.071X_1 + 0.167X_2 + 0.125X_3 + 100 = X_2, \qquad (2)$$

and

$$0.143X_1 + 0.083X_2 + 0.125X_3 + \ \ 45 = X_3. \qquad (3)$$

Writing the system in matrix notation gives

$$\begin{bmatrix} 0.286 & 0.083 & 0.375 \\ 0.071 & 0.167 & 0.125 \\ 0.143 & 0.083 & 0.125 \end{bmatrix} \begin{bmatrix} X_1 \\ X_2 \\ X_3 \end{bmatrix} + \begin{bmatrix} 50 \\ 100 \\ 45 \end{bmatrix} = \begin{bmatrix} X_1 \\ X_2 \\ X_3 \end{bmatrix},$$

$$\mathbf{Ax} + \mathbf{b}_3 = \mathbf{x}, \qquad (4)$$

where $\mathbf{A}$ is the matrix of technical coefficients (Table 9-2).

Multiplying $\mathbf{x}$ by an identity matrix and subtracting the term $\mathbf{Ax}$ from both sides give

$$\mathbf{b}_3 = \mathbf{Ix} - \mathbf{Ax},$$

or

$$(\mathbf{I} - \mathbf{A})\mathbf{x} = \mathbf{b}_3. \qquad (5)$$

Finally, multiplying both sides by the inverse matrix of $(\mathbf{I} - \mathbf{A})$, we obtain the solution, or the total output vector:

$$\mathbf{x} = (\mathbf{I} - \mathbf{A})^{-1}\mathbf{b}_3. \qquad (6)$$

We need to find the actual matrix $(\mathbf{I} - \mathbf{A})^{-1}$ to obtain specific results. The matrix $(\mathbf{I} - \mathbf{A})$ is

$$\begin{bmatrix} 1 & 0 & 0 \\ 0 & 1 & 0 \\ 0 & 0 & 1 \end{bmatrix} - \begin{bmatrix} 0.286 & 0.083 & 0.375 \\ 0.071 & 0.167 & 0.125 \\ 0.143 & 0.083 & 0.125 \end{bmatrix} = \begin{bmatrix} 0.174 & -0.083 & -0.375 \\ -0.071 & 0.833 & -0.125 \\ -0.143 & -0.083 & 0.875 \end{bmatrix}.$$

Applying the Gauss-Jordan procedure to $(\mathbf{I} - \mathbf{A})$ and $\mathbf{I}$ simultaneously, we transform $(\mathbf{I} - \mathbf{A})$ into an identity matrix and $\mathbf{I}$ into $(\mathbf{I} - \mathbf{A})^{-1}$. The result is

$$(\mathbf{I} - \mathbf{A})^{-1} = \begin{bmatrix} 1.564 & 0.226 & 0.702 \\ 0.174 & 1.243 & 0.252 \\ 0.272 & 0.155 & 1.282 \end{bmatrix}.$$

The total output vector can now be calculated by using equation (6):

$$\begin{bmatrix} X_1 \\ X_2 \\ X_3 \end{bmatrix} = \begin{bmatrix} 1.564 & 0.226 & 0.702 \\ 0.174 & 1.243 & 0.252 \\ 0.272 & 0.155 & 1.282 \end{bmatrix} \begin{bmatrix} 50 \\ 100 \\ 45 \end{bmatrix} = \begin{bmatrix} 132.39 \\ 144.34 \\ 86.79 \end{bmatrix}.$$

Since our interest is in finding the expected total output of industry B in three years, we are only concerned with X_2:

$$X_2 = 0.174(50) + 1.243(100) + 0.252(45) = 144.34. \tag{7}$$

Thus, *final* consumer demand for industry B was forecast to increase from \$80 to \$100, or *by 25%*. But, *total* demand for the output of industry B, which takes into account changes in consumer demand in related industries, is expected to increase from \$120 to \$144.34, or *by only 20%*. This is reasonable, since final demand in A was forecast to drop, which would mean A would purchase a smaller quantity of B for raw material. Similarly, final demand in C was forecast to increase, but less rapidly than in B.

If we now wish to vary the *final* demand forecasts for industries A, B, and C, in order to examine the effect on the *total* demand for B's output, we can do so easily.

Let the unknown final demand vector for three years hence be

$$\mathbf{b}_3 = \begin{bmatrix} X_{1F} \\ X_{2F} \\ X_{3F} \end{bmatrix}.$$

Substituting these unknown forecasts into equation (7) yields

$$X_2 = 0.174(X_{1F}) + 1.243(X_{2F}) + 0.252(X_{3F}). \tag{8}$$

Using equation (8) it is easy to examine the effects of changes in final demand in other industries on the total output of industry B, namely X_2. For example, should consumers demand an additional \$10 worth of commodity A, the total output of industry B would need to increase by 0.174×10, or by \$1.74.

CONCLUSION

In this chapter we have described the use of input-output analysis in forecasting total industry output. The basic idea is that goods and services are needed as inputs, or raw materials, in other industries as well as for final demand by consumers. This implies that a single industry cannot be treated in isolation but should be viewed as a part of an interrelated system.

Although input-output analysis has perhaps been most useful to economic—planning groups concerned with an entire economy, it can be a valuable forecasting tool for particular industries and even individual firms. In fact, there is no reason why input-output analysis could not be used by a multi-product firm, where some products serve as inputs to other products and the production processes can be characterized by fixed proportions.

Problems

9-1. Suppose the forecasted final demand for the example in the chapter is

$$\mathbf{b}_3 = \begin{bmatrix} 60 \\ 90 \\ 50 \end{bmatrix}.$$

Find the total outputs in industries A, B, and C.

9-2. Assume that the Households row in Table 9-1 now represents Labor only. Thus, industry A requires 70 units of Labor to produce its total output of 140 units. Is the final demand in Problem 9-1 consistent with a total Labor availability in the economy of 200 units?

9-3. A two-industry economy is characterized by the table of technical coefficients below:

	A	B
A	0	0.5
B	0.5	0

Each unit of A requires 1 unit of labor and 1 unit of capital. Each unit of B requires 2 units of labor. The economy has available resources of 200 units of labor and $66\frac{2}{3}$ units of capital.

Denote the final consumer demand for A and B by D_A and D_B, respectively. Measuring D_A and D_B on the axes, show graphically the feasible region of final demands for the economy. *Hint:* Use equation (5) in the chapter to obtain A and B as functions of D_A and D_B. Use these to eliminate A and B in the resource availability restraints.

9-4. If the economy of Problem 9-3 valued the final demands with prices of \$1 for D_A and \$0.80 for D_B, what would be the optimum output program? What would be the corresponding total outputs of industries A and B? Find a price ratio of D_A and D_B which would make the shadow price, or imputed value, of capital equal to zero.

Game Theory

10

Game theory is the mathematical analysis of situations in which there is a conflict of interest. This relatively new mathematical tool was developed by von Neumann and Morgenstern in 1944.[1]

Examples of game theoretic situations in business are plentiful. Collective bargaining is one. A second is the competition for market share in an oligopolistic market—a market supplied by a small number of firms.

Consider a duopolistic market—a market supplied by two firms, A and B. The management of firm A has a large number of decision variables which affect its profit. Represent these decision variables (price, advertising, output, etc.) by the variables X_i $(i = 1, 2, \ldots, n)$. Similarly, represent firm B's decision variables by Y_i $(i = 1, 2, \ldots, n)$. Then A's profit function, Z_A, can be written as

$$Z_A = F(X_1, X_2, \ldots, X_n; \ Y_1, Y_2, \ldots, Y_n).$$

A's problem is to choose those values of the X_i which will maximize Z_A. However, Z_A is also a function of the decision variables of firm B. What values must A assume for the Y_i?

[1] J. von Neumann and O. Morgenstern, *Theory of Games and Economic Behavior* (Princeton, N.J.: Princeton University Press, 1944).

Conceptually, this problem is fundamentally different from a maximizing problem of the linear programming type. Not only are the values of the Y_i unknown to A, but A must expect B to choose its decision variables in such a way as to tend to diminish Z_A. That is, it is likely that higher profits for A mean lower profits for B. Hence B will probably seek to set prices and to advertise in a way that will increase its profit at the expense of A's profit.

Game theory, unlike linear programming, is not a practical problem solving tool. The reader should think about the problem of estimating the Z_A function to understand why. Nevertheless, the theory of games does have a contribution to make to the study of management and economics; it provides a conceptual framework which permits one to gain an understanding of complex conflict situations. In time, game theory may provide more direct approaches to the solution of business problems.

DEFINITIONS

A *game* is a conflict of interest situation in which two or more *players* participate. Each player has a number of courses of action, or *strategies*, and is assumed to have a known *payoff function*. The payoff can be measured in various ways, such as profit and market share. The payoff function is dependent upon the strategies of all players.

An example payoff function can be written in the form of a *payoff table*. Thus, let A and B each have two strategies, where A's payoff function is that shown below.

A's Payoff Table.

		B's Strategies.	
		1	2
A's Strategies	1	40	10
	2	30	50

The payoffs for the case in which A selects its strategy 1 are given in the first row. Clearly, A's payoff will depend upon which strategy B chooses. If B selects its strategy 1, then A will get a payoff of 40; if B chooses its strategy 2, A will obtain only 10. Similarly, the payoffs to *A* corresponding to its strategy 2 are shown in the second row.

Obviously, if *A knows beforehand* which strategy *B* will play, the problem is relatively simple. Thus, if *B* is expected to play its strategy 2, then *A* must choose either its first strategy and a payoff of 10 or its second strategy with its associated payoff of 50. Clearly, strategy 2 is preferable to *A*. But suppose *B*'s play is not known. If *A* chooses strategy 2 its payoff might be 50, but it might also be 30 (if *B* plays its strategy 1 rather than 2).

Although we have used examples of *two-person* games, there can also be *n-person* games. Since two person games are by far the simplest, we shall limit our discussion to this type. For an introductory discussion of *n*-person games, see W. Baumol, *Economic Theory and Operations Analysis*.

There are a number of other ways to classify games. *Zero-sum* games have the property that the payoff to *A* is the negative of the payoff to *B*, for all outcomes. For example, if the payoff is 50 to *A*, it must be −50 to *B*. In other words, what *A* gains is just exactly what *B* loses, and vice versa. In such a game, the players are said to be in *pure opposition*. Nothing can be gained by *cooperation*. A zero-sum game is a special case of a *constant-sum* game where the sum of the payoffs to the players, for all outcomes, is some constant, but not necessarily zero.

In *nonconstant-sum* games, the sum of the payoffs to the players varies over the outcomes. For example, our duopoly illustration would be a nonconstant-sum game, because the sum of the profits of *A* and *B* would vary as a result of their price and advertising policies. Hence, in games of this type, the players may benefit by cooperation. For further examples of nonconstant-sum games, the reader can refer to Naylor and Vernon, *Microeconomics and Decision Models of the Firm*. Problem 10-4 at the end of this chapter is also an example of a nonconstant-sum game.

We now turn to some simple examples to illustrate the various types of games. Our examples will be of the constant-sum type, and will include both *strictly determined* solutions and *mixed strategy* solutions. The concluding section of this chapter will demonstrate the mathematical relation between game theory and linear programming.

STRICTLY DETERMINED GAMES

Consider the duopoly again, and assume that market share is now the important goal which *A* and *B* seek to maximize. Strictly speaking, we should not refer to them as maximizing—at least not in the usual sense of the word. Firm *A* has the payoff table shown in Table 10-1, where the payoff is expressed in percentage of total market sales.

The strategies refer to various combinations of decision variables. For example, *A*'s strategy 1 might mean a price of $3.98, some small amount of television advertising and heavy direct mail promotion. We permit *B* to have three strategies to emphasize that there is no necessity for a square table.

Which strategy will firm *A* select? Recall that both players are assumed to know the payoff table but *not* what strategy the other player will select.

TABLE 10-1. Firm A's Payoff Table.

		B's Strategies		
		1	2	3
A's Strategies	1	30	25	40
	2	45	20	30

There is no "correct answer" to the question. Firm *A* might decide to choose the strategy with the highest *average* of possible payoffs. Thus, *A* might be indifferent to the choice between 1 and 2, because the average is 31.67 for both. On the other hand, *A* might be a gambler and choose strategy 2 because the highest possible market share, 45, would result if *B* should play its strategy 1.

The game theory answer may now be given. Both firms are assumed to be rather conservative and to "expect the worst." Thus, if *A* were to play strategy 1, it would expect *B* to play that strategy which is most harmful to *A*, namely, *B*'s strategy 2. Notice that the outcome, 25, is the minimum value of all payoffs in the first row. Similarly, if *A* were to play strategy 2, it would expect *B* to play its strategy which would result in the minimum value in row 2, or *B*'s strategy 2.

Firm *A*, therefore, *expects* to get a market share of 25 if it selects strategy 1, and a market share of 20 if it selects strategy 2. That is, *A* expects the minimum of the row corresponding to the strategy. But, which of the strategies will it choose? The answer is now obvious. Firm *A* will select the strategy with the maximum *expected* payoff, or its strategy 1. This is the maximum of the row minima, and is usually called the *maximin* strategy. We show this calculation on the right side of Table 10-2.

The calculation of *B*'s play is shown at the bottom of Table 10-2. This is *B*'s *minimax* strategy. Since this is a constant-sum game, the sum of the payoffs is the

TABLE 10-2. Firm A's Payoff Table.

		B's Strategies 1	2	3	*Minima of Rows*
A's Strategies	1	30	25	40	25 (*Maximin*)
	2	45	20	30	20
Maxima of Columns		45	25 (*Minimax*)	40	

same for all outcomes. Market share will always be 100 percent. Therefore, we could easily construct *B*'s payoff table from *A*'s table. Each entry would be 100 less the corresponding entry in *A*'s table. But this is not necessary. We can see from Table 10-2, which shows the market share going to *A*, that the lower the share, the higher will be *B*'s market share. So *B*, also expecting the worst, expects the payoff to be the highest number possible for each of its strategies. This means that if *B* were to play its strategy 1, it would expect the payoff in Table 10-2 to be 45, the maximum of the payoffs in the first column. Thus, *B expects* the payoff *to A* to be the column maxima for its three strategies. But *B*'s share becomes greater the lower is *A*'s, so *B* will *minimax*, or select that strategy which corresponds to the minimum of the column maxima. This is *B*'s strategy 2. *A* will play strategy 1 and *B* will play strategy 2, the market share going to *A* will be 25 and *B*'s share will be 75.

Notice an important property of this solution: each firm expects the payoff that it, in fact, receives. Hence, neither firm has an incentive to change its play. The reader should check this in Table 10-2. Games with this property are called *strictly determined* games. *A*'s maximin payoff equals *B*'s minimax payoff. Not all constant-sum games have this property, as we shall now show.

MIXED STRATEGY GAMES

To illustrate a nonstrictly determined game, consider the payoff table below (Table 10-3).

Firm *A* would play its maximin strategy 1 and expect a payoff of 30; firm *B* would play its minimax strategy 2 and expect a payoff of 40. The actual payoff would be 40. Notice that *A* is pleasantly surprised—it expected 30 but actually obtained 40.

TABLE 10-3. Firm A's Payoff Table.

		B's Strategies 1	2	*Minima of Rows*
A's Strategies	1	30	40	30 (*Maximin*)
	2	50	20	20
Maxima of Columns		50	40 (*Minimax*)	

Firm B is unhappy with this outcome and has a strong incentive to change its play to strategy 1. The incentive is that, assuming A continues to play its strategy 1, B can reduce A's share from 40 to 30. But it is not reasonable to assume that A will hold fast with strategy 1, because as soon as B switches to strategy 1 A will want to change to its strategy 2. Of course, pursuing this line of argument leads to a nonending series of strategy changes. There does not seem to be a stable solution

By introducing probabilistic decision-making, a stable solution, similar to the strictly determined solution, can be reached. We must assume that the firms do not select a *single* best strategy, but rather choose a mixed strategy of playing each strategy with a certain probability. The concept of mixed strategy can perhaps best be understood if the game is viewed as being repeated over a long period of time. Then, at the beginning of each repetition the firms select the particular strategy for that repetition by a chance device.

Assume that firm A determines that its mixed strategy should be to play strategy 1 with a probability of 1/3 and to play strategy 2 with a probability of 2/3. (We shall describe the determination of the optimal mixed strategy later.) Then, at the beginning of each repetition of the game, firm A selects the particular strategy according to a chance device. For example, firm A could draw a ball at random from a bucket of balls which contains 100 balls bearing the number 1, and 200 balls bearing the number 2. The probability is, therefore, 1/3 that a ball with a 1 on it will be drawn, and 2/3 that a number 2 ball will be drawn. If a number 2 ball is drawn, firm A plays strategy 2.

The problem of firm A is now to find the probability, P, with which it should play its strategy 1. It will, of course, play strategy 2 with a probability of $1 - P$, since the sum of the probabilities must equal unity. Similarly, firm B must determine the probability, Q, with which it should play its strategy 1.

Consider what firm A's payoff will be if B *always* plays its strategy 1. Since A will play its strategy 1 with a probability of P, it is reasonable for A to expect a payoff of 30 (see Table 10-3) a fraction P of the time. Also, A will expect a payoff of 50 a fraction $1 - P$ of the time.

To illustrate, assume that P is 1/3 and $1 - P$ is 2/3. Then, in a large number of repetitions of the game, say 600, A would expect a payoff of 30 about 1/3(600), or 200 times. And, A would expect a payoff of 50 about 2/3(600), or 400 times. The *average* payoff would be

$$\frac{(200)(30)}{600} + \frac{(400)(50)}{600} = 43.33.$$

Of course, in 600 actual repetitions the payoffs of 30 might, by chance, number only 197, even though the true probability of selecting strategy 1 is 1/3. Nevertheless, as the number of repetitions increases the average payoff would tend closer and closer to the theoretically expected average payoff, which can be calculated as

$$P(30) + (1 - P)(50)$$

or, for $P = 1/3$,

$$\frac{1}{3}(30) + \frac{2}{3}(50) = 43.33.$$

In statistics this is referred to as the *expected value*, or *mathematical expectation* of the payoff.

Next, denote the theoretically expected payoff to firm A by EA. To indicate that it is the expected payoff to A *when firm B plays its strategy 1 each time*, we attach the subscript 1, or EA_1. Similarly, EA_2 is A's expected payoff when B only plays its strategy 2. We shall consider the case where B *also* plays a mixed strategy later.

Taking P to be an unknown variable, we can now write expressions for EA_1 and EA_2 (see Table 10-3).

$$EA_1 = P(30) + (1 - P)50, \tag{1}$$

$$EA_2 = P(40) + (1 - P)20. \tag{2}$$

A's problem is to select the value of P which maximizes its minimum expected payoff. In effect, this means that A calculates the minimum expected payoff for every possible value of P, then chooses that P corresponding to the maximum of the minimum expected payoffs. To clarify the problem, we plot equations (1) and (2) in Figure 10-1.

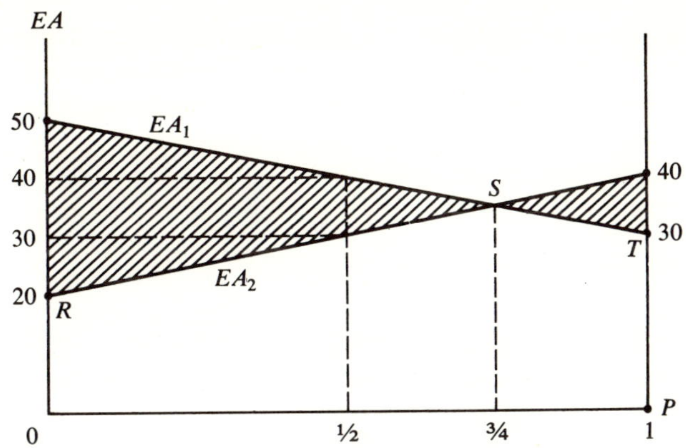

Figure 10-1. Expected Payoffs to A.

The probability, P, with which A plays its strategy 1, is measured horizontally. The vertical axis measures the expected payoff, EA. The probability, P, of course, can range only from 0 to 1. The two straight lines EA_1 and EA_2 were interpreted earlier. Consider the values of EA_1 and EA_2 for $P = 1/2$:

$$EA_1 = \frac{1}{2}(30) + \frac{1}{2}(50) = 40,$$

$$EA_2 = \frac{1}{2}(40) + \frac{1}{2}(20) = 30.$$

In Figure 10-1, the vertical line at $P = 1/2$ intersects EA_1 at a value of 40 and EA_2 at a value of 30. That is, if firm A should play strategy 1 with a probability of 1/2 and strategy 2 with a probability of 1/2, its expected payoff could be as high as 40 (if B played its strategy 1 exclusively), and as low as 30 (if B played only its strategy 2). What about payoffs in between 30 and 40? These payoffs could be expected if B also adopted a mixed strategy. Thus, EA_1 and EA_2 set the extremes; payoffs in between result from B's play of a mixed strategy. The shaded region in Figure 10-1 represents all possible payoffs to A.

Since A is a follower of the maximin strategy, it always expects the worst. Hence, for each value of P in Figure 10-1, A expects the lowest or minimum expected payoff. The lower boundary of the shaded area, the kinked line RST, is the set of all minimum expected payoffs to A. Of these minimum expected payoffs, the payoff at S is the maximum. Hence, A selects the value of P associated with the payoff at S, or $P = 3/4$.

Notice that the value of P and EA can be determined by solving the equations of the intersecting lines, EA_1 and EA_2, simultaneously. The simultaneous solution of equations (1) and (2) is:

$$EA = 35,$$

$$P = \frac{3}{4}.$$

We can also construct a figure displaying B's expected payoffs. Using notation analogous to that used for A, we can write the two equations (see Table 10-3):

$$EB_1 = Q(30) + (1 - Q)40, \qquad (3)$$

$$EB_2 = Q(50) + (1 - Q)20. \qquad (4)$$

Recall that EB_1 is the expected payoff of B expressed as a function of the probability, Q, with which B plays its strategy 1. The subscript 1 means that A is playing its strategy 1 exclusively.

Firm B also expects the worst. But the worst to B is the upper boundary, UVW, in Figure 10-2. The payoffs are all measured in terms of A's market share. Since B's share becomes lower the higher A's share becomes, firm B views high values of EB as bad. Clearly, the minimum of all maximum expected payoffs corresponds to point V in the figure. This point is an intersection point also, and we can determine

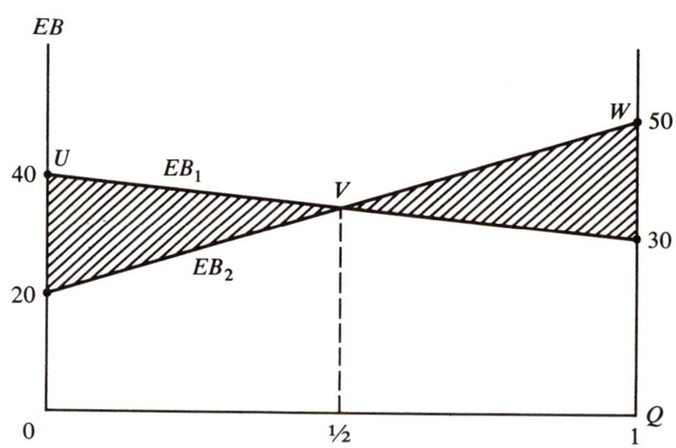

Figure 10-2. Expected Payoffs of B.

the values of P and EB by solving the two equations simultaneously. The solution is

$$EB = 35,$$

$$Q = \frac{1}{2}.$$

Firm A's maximin strategy yields the same expected payoff as does firm B's minimax strategy. Hence, by introducing mixed strategies we have converted a non-strictly determined game into a game with a stable solution. The firms have no incentive to change the probabilities with which they play their strategies.

LINEAR PROGRAMMING RELATION TO MIXED STRATEGY GAMES

Solving mixed strategy games graphically can become difficult if the number of strategies becomes large. It is interesting to observe that mixed strategy games can be cast in a linear programming structure and thereby be solved by the simplex method.

Consider the problem of firm A in our previous example. The firm seeks to maximize EA, subject to three restraints. The first two restraints are derived from equations (1) and (2):

$$P(30) + (1 - P)50 \geq EA, \tag{5}$$

$$P(40) + (1 - P)20 \geq EA. \tag{6}$$

Although EA is a variable to be determined in the problem, restraints (5) and (6) require that the expected payoffs, EA and EA_2, be at least as great as EA. Since the objective function is to *maximize* the variable EA, the object is to maximize the minimum of the possible expected payoffs.

To understand restraints (5) and (6), it may be helpful to refer back to Figure 10-1. Imagine the variable EA as a horizontal line somewhere below the point R. For this value of EA, both EA_1 and EA_2 exceed EA for all values of P. Thus, restraints (5) and (6) are satisfied. However, the objective is to maximize EA. Therefore, imagine pushing the horizontal line upward as far as possible, making sure that, at least for some value of P, restraints (5) and (6) are not violated. Clearly, the horizontal line cannot be pushed any higher once it passes through point S, or the restraints (5) and (6) will be violated. That is, there is no value of P for which both EA_1 and EA_2 are at least as great as EA, if the horizontal line (EA) lies above point S.

A further restraint is needed to ensure that the sum of the probabilities is unity.

In the usual linear programming format, where we use P_1 for P and P_2 for $1 - P$, we get:

Maximize

$$Z = 0P_1 + 0P_2 + 1EA \tag{7}$$

subject to

$$-30P_1 - 50P_2 + 1EA \leq 0, \tag{8}$$

$$-40P_1 - 20P_2 + 1EA \leq 0, \tag{9}$$

$$1P_1 + 1P_2 + 0EA = 1, \tag{10}$$

and

$$P_1, P_2, EA \geq 0. \tag{11}$$

The above problem can be considered the primal problem. If so, what is the dual problem? As the reader may have already suspected, the dual problem is none other than the problem faced by firm *B*. It is no coincidence in our mixed strategy example that the expected payoffs in the solution are the same for *A* as for *B*. This result is assured by Duality Theorem 1 of Chapter 6. The reader is urged to formulate firm *B*'s problem as a linear programming problem in order to check this assertion.

Problems

10-1. Find the solution to the following game:

A's Payoff Table.

		B's Strategies 1	2
A's Strategies	1	50	40
	2	60	20

10-2. Examine the solution to Problem 10-1 graphically, assuming that the players use mixed strategies.

10-3. Solve firm *B*'s problem, in the mixed strategy example of the chapter, by the simplex method.

10-4. The Prisoner's Dilemma is a classic illustration of a nonconstant-sum game. Two apprehended thieves are interrogated separately and each has two strategies: (1) confess, or (2) not confess. The payoff tables of the two thieves, X and Y, are given below, where the punishments in numbers of years in prison are the entries in the tables. (By prefixing minus signs the usual interpretation of payoff tables is still applicable: higher algebraic values are more desirable than lower ones.)

X's Payoff Table.

		Strategies of Y	
		Confess	Not confess
Strategies of X	Confess	−25	0
	Not confess	−100	−5

Y's Payoff Table.

		Strategies of Y	
		Confess	Not confess
Strategies of X	Confess	−25	−100
	Not confess	0	−5

(a) Why is this a nonconstant-sum game?

(b) For each thief, one strategy is preferable to the other *regardless* of the other thief's play. (The preferred strategy is said to *dominate* the other.) Which are the dominant strategies of the two thieves?

(c) If each thief plays his dominant strategy, what are the payoffs?

(d) The payoffs in (c) are less desirable than if each thief elected to not confess. That is, cooperation could lead to a better payoff for each. What reasons can you give to explain why cooperation might be difficult? Thus, explain the dilemma.

(e) Can you think of any business situations which are analogous to the "Prisoner's Dilemma"?

Bibliography

We have indicated by an asterisk (*) those items in each of the categories which should be most useful as supplementary references for the material in this book.

General

* Baumol, William J. *Economic Theory and Operations Analysis.* Englewood Cliffs, New Jersey: Prentice-Hall, 1965.

Bierman, H., C. P. Bonini, and W. H. Hausman. *Quantitative Analysis for Business Decisions.* Homewood, Illinois: Richard D. Irwin, 1969.

* Dorfman, R., P. A. Samuelson, and R. M. Solow. *Linear Programming and Economic Analysis.* New York: McGraw-Hill Book Co., 1958.

Gale, David. *The Theory of Linear Economic Models.* New York: McGraw-Hill Book Co., 1960.

Hillier, F. S., and G. J. Lieberman. *Introduction to Operations Research.* San Francisco: Holden-Day, 1967.

* Naylor, T. H., and J. M. Vernon. *Microeconomics and Decision Models of the Firm.* New York: Harcourt, Brace, and World, 1969.

Mathematics

* Chiang, Alpha C. *Fundamental Methods of Mathematical Economics.* New York: McGraw-Hill Book Co., 1967.

Howell, J. E., and D. Teichroew. *Mathematical Analysis for Business Decisions.* Homewood, Illinois: Richard D. Irwin, 1963.

Teichroew, D. *An Introduction to Management Science.* New York: John Wiley, 1964.

Linear Programming

Baumol, William J. "Activity Analysis in One Lesson." *American Economic Review* XLVIII (December 1958), 837–873.

Charnes, A., W. W. Cooper, and A. Henderson. *An Introduction to Linear Programming.* New York: John Wiley, 1953.

Charnes, A., and W. W. Cooper. *Management Models and Industrial Applications of Linear Programming.* New York: John Wiley, 1961.

Chung, An-Min. *Linear Programming.* Columbus, Ohio: Charles E. Merrill Books, 1959.

Dano, Sven. *Linear Programming in Industry. Theory and Applications.* Vienna: Springer–Verlag, 1960.

Dantzig, George B. *A Procedure for Maximizing a Linear Function Subject to Linear Inequalities.* Washington: Headquarters, U.S. Air Force, Comptroller, 1948.

* Dantzig, George B. *Linear Programming and Extensions.* Princeton: Princeton University Press, 1963.

Dorfman, Robert. "Mathematical, or Linear, Programming: A Nonmathematical Exposition," *American Economic Review* XLIII (December 1953), 797–825.

Garvin, Walter W. *Introduction to Linear Programming.* New York: McGraw-Hill Book Co., 1960.

Gass, Saul. *Linear Programming: Methods and Applications.* New York: McGraw-Hill Book Co., 1964.

* Hadley, G. *Linear Programming.* Reading, Massachusetts: Addison-Wesley Publishing Co., 1962.

Levin, R. I., and R. P. Lamone. *Linear Programming for Management Decisions.* Homewood, Illinois: Richard D. Irwin, 1969.

Llewellyn, R. W. *Linear Programming.* New York: Holt, Rinehart, and Winston, 1964.

Simonnard, Michel. *Linear Programming.* Englewood Cliffs, New Jersey: Prentice-Hall, 1966.

* Spivey, W. Allen. *Linear Programming.* New York: Macmillan Co., 1963.

Vajda, S. *Readings in Linear Programming.* New York: John Wiley, 1958.

Vajda, S. *Mathematical Programming.* Reading, Massachusetts: Addison-Wesley Publishing Co., 1961.

Integer Programming

Balinski, M. L. "Integer Programming: Methods, Uses, Computation," *Management Science*, XII (November 1965), 253–313.

Dantzig, George B. "Note on Solving Linear Programs in Integers," *Naval Research Logistics Quarterly*, VI (1959), 75–76.

Dantzig, George B. "On the Significance of Solving Linear Programming Problems with Some Integer Variables," *Econometrica*, XXVIII (January 1960), 30–44.

Driebeek, Norman J. "An Algorithm for the Solution of Mixed Integer Programming Problems," *Management Science*, XIII (March 1966), 576–587.

Gomory, Ralph E. "Outline of an Algorithm for Integer Solutions to Linear Programs," *Bulletin of the American Mathematical Society*, LXIV (September 1958), 275–278.

Gomory, Ralph E. *An Algorithm for the Mixed Integer Problem*, RM–2597, The Rand Corporation, 1960.

* Hadley, G. *Nonlinear and Dynamic Programming*. Reading, Massachusetts: Addison-Wesley Publishing Co., 1964.

Kendrick, David A. *Programming Investment in the Process Industries*. Cambridge, Mass: Technology Press, 1967.

Lawler, E. L., and D. E. Wood. "Branch and Bound Methods, A Survey," *Operations Research* XIV (July–August, 1966).

* Weingartner, H. M. *Mathematical Programming and the Analysis of Capital Budgeting Problems*. Englewood Cliffs, New Jersey: Prentice-Hall, 1963.

Input–Output Analysis

Almon, Clopper, Jr. "Consistent Forecasting in a Dynamic Multi-Sector Model," *The Review of Economics and Statistics*, May 1963, 148–162.

* Chenery, H. B., and P. G. Clark. *Interindustry Economics*. New York: John Wiley, 1959.

Cleland, Sherril. "Local Input–Output Analysis: A New Business Tool," *Business Topics*, Michigan State University, Autumn, 1959, 41–48.

Evans, W. D., and M. Hoffenberg. "The Interindustry Relations Study for 1947," *The Review of Economics and Statistics*, May 1952, 97–142.

* Evans, W. D. "Marketing Uses of Input–Output Data," *Journal of Marketing*, July 1952, 11–21.

Leontief, W. W. *The Structure of the American Economy, 1919–1939*. New York: Oxford University Press, 1951.

* Miernyk, William H. *The Elements of Input–Output Analysis*. New York: Random House, 1967.

National Bureau of Economic Research. *Input–Output Analysis: An Appraisal*. Princeton: Princeton University Press, 1955.

Yan, Chiou–Shuang. *Introduction to Input–Output Economics*. New York: Holt, Rinehart, and Winston, 1969.

Game Theory

Hurwicz, Leonid. "The Theory of Economic Behavior," *American Economic Review*, XXXV (1945), 909–925.

* Luce, R. D., and H. Raiffa. *Games and Decisions.* New York: John Wiley, 1957.

* McKinsey, J. C. C. *Introduction to the Theory of Games.* New York: McGraw-Hill Book Co., 1952.

Shubik, Martin. *Strategy and Market Structure.* New York: John Wiley, 1959.

Shubik, Martin. *Game Theory and Related Approaches to Social Behavior.* New York: John Wiley, 1964.

von Neumann, J., and O. Morgenstern. *Theory of Games and Economic Behavior*, 2d ed. Princeton, New Jersey: Princeton University Press, 1947.

Williams, J. D. *The Complete Strategist.* New York: McGraw-Hill Book Co., 1954.

Cases in Linear Programming

Introduction to Cases

Each of the fifteen cases included here requires the use of linear programming to determine the final solution. The majority of the cases are based on actual experience and represent a cross section of linear programming applications with regard both to the type of application and to the degree of complexity. Although an electronic computer is not necessary for solving these cases, most of the cases (especially Cases 5–13) are of sufficient magnitude that solutions by manual techniques would require an excessive amount of time.

It is not necessary for the reader to be familiar with computer programming to solve these cases on an electronic computer. Most computers presently being used by industry and colleges and universities are equipped with linear programming codes. These codes are essentially instructions, written in computer language, which direct the computer to apply the simplex method (or the revised simplex method). Hence, to solve a linear programming problem on the computer, the reader needs only to write out the objective function and restraints in a format prescribed by the instructions for the code. For example, the code might specify that in card punching, symbols for variables must be followed by coefficients and constants in a particular order.

To illustrate the speed of an electronic computer solution, the Mathematical Programming System of the I. B. M. 360/75 computer was used to solve The Wilcox

Manufacturing Case in 6.6 seconds! It required 24 equations in 72 variables. A total of 32 iterations were needed before obtaining the optimum solution.

Even if a computer is not available, you can derive the following major benefits from these cases: (1) practice in recognizing real-life linear programming problems, and (2) practice in formulating industrial problems as linear programming problems. In fact, the reader should not attempt to solve any of these problems completely unless he has access to a computer, as some of the problems require over 100 iterations to obtain an optimum solution. However, each problem should be formulated mathematically as a linear programming model.

Cases 1 and 2 (The Evangeline Coffee Company and the Shenandoah Valley Textile Mill) have been solved completely, and the solutions appear at the end of each of these cases. The remaining thirteen cases are to be solved by the reader. Case 3 (The Jones Company) should be solved graphically. Cases 4–13 are larger cases, all of which can be solved by the simplex method described in Chapter 5. Case 14 (Maluland) is an example of input-output analysis where choice among final demands is possible. That is, the Maluland Case should be converted into a linear programming problem. Chapter 9 contains the necessary background. The last case, Case 15 (The Key Largo Shipyard), requires a knowledge of game theory, which can be found in Chapter 10.

The questions included at the end of each case relate not only to the linear programming technique itself, but also to some of the managerial considerations involved, such as defining the problem, collecting pertinent data, solving the problem with the aid of a computer, and implementing the solution. In each case, both the name of the company and the pertinent data have been disguised to preserve the anonymity of the firms involved.

The Evangeline Coffee Company

Cas

The Evangeline Coffee Company of New Orleans, Louisiana, imports, blends, and roasts green coffees[1] for distribution in a six-state area along the Gulf Coast. The company markets three different blends of coffee under the following brand names: Cafe d'Elite, Plantation, and Creole. Cafe d'Elite is a luxury item served exclusively by leading hotels, fine restaurants, and espresso coffee houses. Plantation is the company's leading product in terms of volume. It has widespread distribution throughout the company's six-state marketing area. The Creole blend is a dark roast, low quality coffee with chicory[2] that is sold only in southern Louisiana.

COFFEE BLENDING

Coffee is by no means a homogeneous commodity. A beverage may be made from any coffee, but the product sold commercially to the ultimate consumer is usually a blend. Blending is necessary to harmonize the widely varying characteristics

[1] Coffees that have not been roasted are referred to as "green coffees" by the trade.

[2] Chicory is a brownish carrot-like root, available in large quantities at a negligible cost. It is added in a fixed proportion by weight to coffees sold in southern Louisiana. For the purposes of this study it may be ignored.

of different coffees in order to produce a desired flavor. Coffees differ in inherent characteristics and are further differentiated on a quality basis by the prominence of such characteristics. Thus, the problem of the trade is to translate consumer taste distinctions into the kinds and proportions of coffees forming a blend necessary to produce a given result in the cup.

The seven most important characteristics of coffees and blends of coffee have been found to be strength, acidity, caffeine content, liquoring value, hardness, aroma sweetness, and body. All of these characteristics are described by a trade jargon that helps to identify any particular lot as to its commercial utility, and therefore more or less establishes a value for it in relation to all other kinds, types, and grades of coffees.

The composition of any given blend of coffee is usually a closely guarded trade secret. Standard good brands of coffee in the United States are a blend of 15 to 40 percent of mild Columbian or Central American coffees with good Brazilian coffees. High-quality brands are blends of milds and may contain no Brazilian coffee, whereas the cheapest coffees may consist entirely of Brazils.

Coffee merchants have considerable latitude in varying their blend formulas as conditions dictate. Changing price differentials between coffee types may require changes in formulas from month to month. Generally, such changes will be undertaken for the purpose of adjusting cost factors without creating any noticeable change in cup quality. Cup testing, performed by expert coffee tasters, is essential to arriving at just the right combination of coffee.

RECENT DEVELOPMENTS

As a result of a recent operations research study, Evangeline has achieved considerable success in attempting to quantify the 7 major taste characteristics of coffee. This was accomplished by the development of a rating scale for 5 of the taste attributes of coffee. For example, every type of green coffee that is available in any significant quantity has been assigned an index number between 0 and 10 according to its relative "strength." Those coffees that have a "very strong" taste are assigned numbers close to 10. Those that are relatively mild are assigned numbers near 5. Weak coffees are assigned numbers near 0. Similar indices have been developed to express relative differences in liquoring value, hardness, aroma, and body. It was not necessary to develop such an index to measure acidity or caffeine content, for acidity can be determined by measuring the *p*H factor, and caffeine content is measured as a percent of weight. It has also been found that the taste characteristics of different types of coffees combine *linearly* when blended.

EVANGELINE'S BLENDS

The Evangeline Coffee Company has developed a rigid set of requirements for each of its 3 blends. A failure to satisfy these requirements may result in a significant decline in sales. Hence, every possible step must be taken to ensure blends of constant quality within the limits stipulated below.

Cafe d'Elite (*high quality*)		*Plantation* (*medium quality*)		*Creole* (*low quality*)	
Strength	≤ 8.0	Strength	≤ 7.0	Strength	≥ 9.0
Acidity	≤ 3.5	Acidity	≤ 4.0	Acidity	≤ 5.0
Caffeine	≤ 2.8	Caffeine	≤ 2.2	Caffeine	≤ 2.4
Liquoring value	≥ 7.0	Liquoring value	≥ 6.0	Liquoring value	≥ 5.0
Hardness	≤ 2.5	Hardness	≤ 3.0	Hardness	≤ 7.8
Aroma	≥ 7.0	Aroma	≥ 5.0	Aroma	≥ 4.0
Body	≥ 7.0	Body	≥ 6.0	Body	≥ 7.0

On the first day of each month the company forecasts its green coffee requirements for the period of time beginning one month from that day. (The company operates on the basis of a monthly planning period.) Since green coffees are imported from Latin America and Africa, it may take as long as one month from the time an order is placed before it arrives at the Port of New Orleans. Hence, one month of lead time is required for all orders. For example, green coffees to be roasted in February must be ordered on the first day of January.

On the first day of a particular month the company found that its available supply of green coffees was limited to 12 different types as indicated in Table I. These 12 coffees could be delivered to New Orleans before the first day of the following month. According to Table I, these coffees vary according to 1. price, 2. quantity available, and 3. taste characteristics.

Forecasts indicate that the demand for the company's 3 blends during the coming month will be as follows.

Blend	*Forecast Demand*
Cafe d'Elite	10,000 pounds
Plantation	60,000 pounds
Creole[3]	30,000 pounds

[3] Chicory will be added to this quantity in a fixed proportion by weight. It will not affect the linear relationship that exists among the taste factors of the blend.

TABLE I

Green Coffees Available	Price per Pound	Available Supply in Pounds	Strength Index	Acidity pH	Percent Caffeine Content	Liquoring Value	Hardness Index	Aroma Sweetness Index	Body Index
1. Brazils									
Santos 4's	$0.35	25,000	6	4.0	1.8	6	2	8	8
Bourbon Santos	0.36	10,000	6	3.9	1.6	6	3	9	8
Rios	0.20	75,000	10	4.5	1.0	5	7	4	6
Victorias	0.17	50,000	10	5.0	0.9	3	8	2	6
2. Columbians									
Medellins	$0.44	5,000	8	3.0	3.0	8	2	7	9
Bogotas	0.43	5,000	9	3.0	2.5	7	2	6	8
Manizales	0.45	4,000	8	3.0	3.0	9	1	7	10
Bucara-Mangas	0.41	20,000	7	4.0	2.0	6	2	7	7
3. Mexicans	$0.36	5,000	5	3.5	1.5	6	3	9	8
4. Guatemalans	$0.34	4,000	6	3.6	1.1	6	4	7	7
5. El Salvadors	$0.36	5,000	6	3.2	1.4	6	3	8	8
6. Africans									
Robustas	$0.19	100,000	10	5.1	1.7	5	9	1	5

The company is confronted with the problem of determining an optimum combination of available green coffees for next month's roasting operation. It must determine a combination that will satisfy both the demand requirements and quality requirements for each of its 3 blends as well as minimize the total cost of green coffee purchases.

SOLUTION

Since the formulation of this problem as a linear programming model consists of 36 equations and 82 unknowns (including slack and artificial variables) and requires 72 iterations to reach an optimum solution, a complete presentation of all of the linear restraints and simplex tables involved in solving the problem cannot possibly be included in a book of this size. Furthermore, it is highly unlikely that anyone would even consider solving a problem of this magnitude by hand. Hence, we shall consider only a concise mathematical formulation of the problem with examples of the different types of restraints involved. The mathematical treatment of the formulation of this problem is of a form that is usually found in the literature on linear programming. The reader is encouraged to complete the formulation of the problem, set up an initial simplex table, and solve the problem on an electronic computer. The final answer to the problem is included at the end of the following explanation.

Definition of Variables

Let X_{ij} = the quantity of the ith green coffee used in the jth blend, where ($i = 1, 2, \ldots, 12$) and ($j = 1, 2, 3$).

Statement of Objective Function

Let C_i = the cost per pound of the ith coffee ($i = 1, \ldots, 12$). Then the objective of the coffee company is to minimize

$$\sum_{i=1}^{12} \sum_{j=1}^{3} C_i X_{ij}.$$

This is equivalent to saying the company wants to minimize the function

$$\begin{aligned} Z = {} & 35X_{11} + 36X_{21} + 20X_{31} + 17X_{41} + 44X_{51} + 43X_{61} \\ & + 45X_{71} + 41X_{81} + 36X_{91} + 34X_{10,1} + 36X_{11,1} + 19X_{12,1} \\ & + 35X_{12} + 36X_{22} + 20X_{32} + 17X_{42} + 44X_{52} + 43X_{62} \\ & + 45X_{72} + 41X_{82} + 36X_{92} + 34X_{10,2} + 36X_{11,2} + 19X_{12,2} \\ & + 35X_{13} + 36X_{23} + 20X_{33} + 17X_{43} + 44X_{53} + 43X_{63} \\ & + 45X_{73} + 41X_{83} + 36X_{93} + 34X_{10,3} + 36X_{11,3} + 19X_{12,3}. \end{aligned}$$

Formulation of Linear Restraints

There are 3 different types of linear restraints imposed on the variables (X_{ij}'s) in this problem: demand, supply, and quality.

Demand Restraints. There are 3 demand restraints, one for each of the company's products. These may be represented by the mathematical expression

$$\sum_{i=1}^{12} X_{ij} = D_j \qquad j = 1, 2, 3,$$

where D_j is the demand for the jth blend ($j = 1, 2, 3$). For example the demand restraint for Cafe d'Elite (where $j = 1$) would be

$$X_{11} + X_{21} + X_{31} + X_{41} + X_{51} + X_{61} + X_{71} + X_{81} + X_{91} + X_{10,1} + X_{11,1} + X_{12,1} = 10{,}000.$$

The demand restraints for the Plantation and Creole blends would be of a similar nature.

Supply Restraints. There are 12 supply restraints, one for each of the 12 available green coffees. These may be represented by the expression

$$\sum_{j=1}^{3} X_{ij} \leq S_i \quad i = 1, 2, \ldots, 12,$$

where S_i is the available supply of the ith green coffee. For example, the supply restraint for Santos 4's (where $i = 1$) would be expressed as

$$X_{11} + X_{12} + X_{13} \leq 25{,}000.$$

The supply restraints for the other green coffees would be of a similar form.

Quality Restraints. There are 21 quality restraints, 7 for each of the company's 3 blends. These may be expressed as

$$\sum_{i=1}^{12} a_{ik} X_{ij} \begin{pmatrix} \leq \\ = \\ \geq \end{pmatrix} D_j b_{jk}, \qquad \begin{matrix} j = 1, 2, 3, \\ k = 1, 2, \ldots, 7, \end{matrix}$$

where

a_{ik} = the kth quality characteristic of the ith green coffee;
D_j = the demand for the jth blend;
b_{jk} = the kth characteristic of the jth blend.

For example, the "strength" restriction for Cafe d'Elite (where $j = 1$ and $k = 1$) may be expressed as

$$6X_{11} + 6X_{21} + 10X_{31} + 10X_{41} + 8X_{51} + 9X_{61} + 8X_{71} + 7X_{81} + 5X_{91} + 6X_{10,1} + 6X_{11,1} + 10X_{12,1} \leq (10{,}000)(8).$$

Similar restraints must be included for each of the 7 quality characteristics of the 3 blends.

Final Solution

After the 36 restraints have been formulated and slack and artificial variables have been introduced, a basic feasible solution and an initial simplex table may be developed. The simplex table of this problem consists of 36 equations and 82 unknowns (36 real variables and 48 slack and artificial variables). The optimum solution resulted in a total expenditure of $33,558.11 for green coffees. The coffees included in the optimum solutions are indicated in Table 2.

TABLE 2

Green Coffee	*Blend*		
	Cafe d'Elite	Plantation	Creole
Santos 4's		20,000	5,000
Bourbon Santos		10,000	
Rios			5,000
Victorias			12,000
Medellins	2,750	2,250	
Bogotas			5,000
Manizales	1,500		2,500
Bucara-Mangas		15,000	
Mexicans		5,000	
Guatemalans	750	3,250	
El Salvadors	5,000		
Robustas		4,500	

The Shenandoah Valley Textile Mill *Case 2*

The Shenandoah Valley Textile Mill purchases large quantities of rough, unfinished cotton cloth from nearby weaving mills and through a series of finishing operations converts this cloth into a wide variety of lusters, colors, textures, and patterns. Through these finishing operations it is also possible to achieve starched, preshrunk, glazed, wrinkle-resistant, napped, and a large number of other special finishes.

Finishing operations may be divided into three major groups: preparatory, color, and finishing. Preparatory operations include laying out, singeing, desizing, kier boiling, bleaching, and drying. Color operations include printing and dyeing. Finishing operations include starching, shrinking, calendering, and napping, as well as other special finishing processes. The process flow chart for the Shenandoah Valley Mill is outlined in Figure 1.

The marketing research department has recommended the production of 7 different styles of cotton cloth for the forthcoming week: 1 bleached style (B); 4 printed styles (P_1, P_2, P_3, and P_4); and 2 dyed styles, blue and red (D_1 and D_2). The company is committed by a contract with a garment factory to produce at least 5000

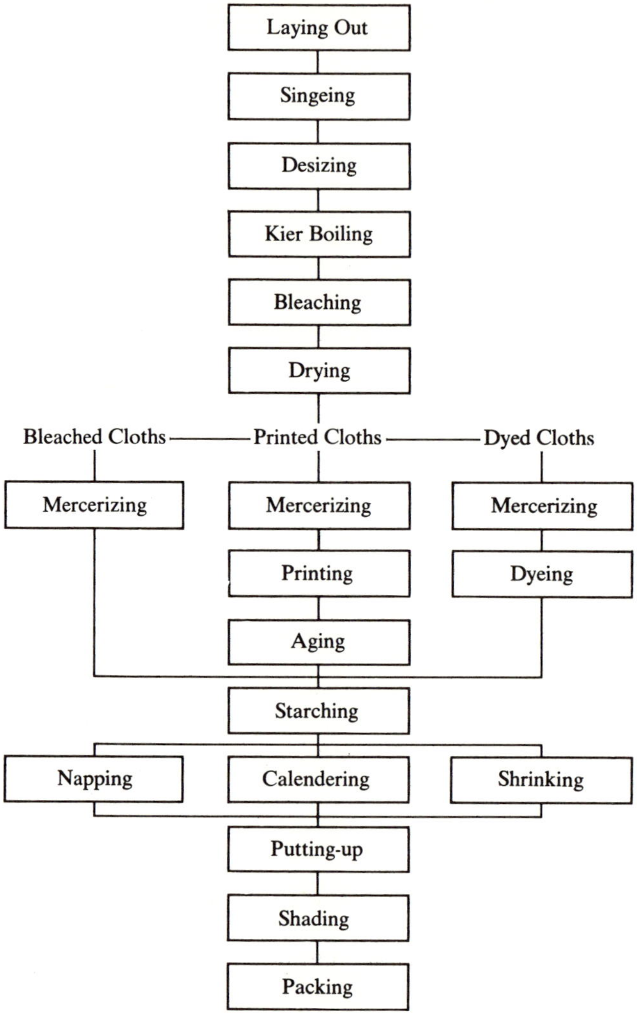

Figure 1. Process Flow Chart for the Shenandoah Valley Textile Mill.*

* This flow chart was obtained from Stanley S. Miller, et al., *Manufacturing Policy* (Homewood, Ill.: Richard D. Irwin, 1957), p. 162. Reprinted by permission.

yards of each of the printed styles in the following week. The sales department has estimated that the maximum possible sales of styles P_3 and P_4 for the forthcoming week will be 100,000 yards and 50,000 yards respectively. Barring any unforeseen developments in marketing conditions, management estimates that it can sell all that it produces of the other 5 styles.

If the mill produces at its maximum capacity next week it is certain that there will be excess capacity in at least 6 of its operations: laying out, napping, shrinking, putting-up, shading, and packing. Hence, these operations are assumed to be neither restrictive nor critical and therefore, may be omitted from further consideration.

Production rates for each of the 7 styles for each of the 11 restrictive operations are given below. These rates are expressed in yards per hour.

	Styles						
Process	*B*	P_1	P_2	P_3	P_4	D_1	D_2
Singeing	9,000	6,000	9,000	7,000	8,000	9,000	8,000
Desizing	13,000	10,000	9,000	11,000	8,000	13,000	12,000
Kier Boiling	1,500	900	1,000	800	900	1,300	1,200
Bleaching	1,000	1,100	1,050	1,100	1,100	1,100	1,200
Drying	13,000	10,000	10,000	12,000	11,000	11,000	12,000
Mercerizing	800	550	600	650	700	700	800
Printing	—	300	300	200	250	—	—
Aging	—	5,000	4,000	4,000	6,000	—	—
Dyeing	—	—	—	—	—	—	—
(blue)	—	—	—	—	—	4,000	—
(red)	—	—	—	—	—	—	3,500
Starching	2,000	1,800	1,800	1,600	1,500	2,000	1,500
Calendering	4,000	5,000	3,000	2,500	4,000	3,200	3,500

The maximum number of process-hours that will be available next week for each critical operation are shown overleaf. Although the mill will be in operation 3 shifts per day, for 6 days during the forthcoming week, in estimating process capacities allowances have been made for breakdowns, maintenance, cleaning, etc. (Figures are shown in process-hours.)

Process	*Capacity*
Singeing	150
Desizing	150
Kier Boiling	900
Bleaching	1500
Drying	140
Mercerizing	
(bleached)	830
(printed)	830
(dyed)	830
Printing	1800
Aging	150
Dyeing	
(blue)	150
(red)	140
Starching	500
Calendering	450

Determine the most profitable production schedule for the following week if the contributions to profit and overhead per yard for each style are as follows.

Style	*Profitability*
B	\$0.40
P_1	0.60
P_2	0.80
P_3	1.00
P_4	1.25
D_1	1.20
D_2	1.30

SOLUTION

Definition of Variables

Let B, P_1, P_2, P_3, P_4, D_1, and D_2 be the production quantities of each of the 7 products.

Statement of Objective Function

Then the objective function may be expressed as

$$\text{(maximize)} \quad Z = 0.40B + 0.60P_1 + 0.80P_2 + 1.00P_3 + 1.25P_4 + 1.20D_1 + 1.30D_2$$

Formulation of Linear Restraints

There are 2 different sets of linear restraints in this problem: 14 process restraints and 6 demand restraints. Since the process restraints are expressed in process-hours, it is necessary to use the reciprocals of the production rates (hours per yard) as coefficients rather than yards per hour. For example, the first process restraint is

$$\frac{B}{9000} + \frac{P_1}{6000} + \frac{P_2}{9000} + \frac{P_3}{7000} + \frac{P_4}{8000} + \frac{D_1}{9000} + \frac{D_2}{8000} \leq 150,$$

or

$$1.1B + 1.7P_1 + 1.1P_2 + 1.4P_3 + 1.2P_4 + 1.1D_1 + 1.2D_2 \leq 1{,}500{,}000.$$

The entire sets of process restraints and demand restraints are given below.

Process Restraints. The process restraints to be considered are:

Singeing	$1.1B + 1.7P_1 + 1.1P_2 + 1.4P_3 + 1.2P_4 + 1.1D_1 + 1.2D_2 \leq 1{,}500{,}000,$
Desizing	$7.7B + 10.0P_1 + 11.1P_2 + 9.1P_3 + 12.5P_4 + 7.7D_1 + 8.3D_2 \leq 15{,}000{,}000,$
Kier Boiling	$6.7B + 11.1P_1 + 10.0P_2 + 12.5P_3 + 11.1P_4 + 7.7D_1 + 8.3D_2 \leq 9{,}000{,}000,$
Bleaching	$10.0B + 9.1P_1 + 9.5P_2 + 9.1P_3 + 9.1P_4 + 9.1D_1 + 8.3D_2 \leq 15{,}000{,}000,$
Drying	$7.7B + 10.0P_1 + 10.0P_2 + 8.3P_3 + 9.1P_4 + 9.1D_1 + 8.3D_2 \leq 14{,}000{,}000,$
Mercerizing	
(bleached)	$1.2B + \quad \leq 830{,}000,$
(printed)	$1.8P_1 + 1.7P_2 + 1.5P_3 + 1.4P_4 \leq 830{,}000,$
(dyed)	$1.4D_1 + 1.2D_2 \leq 830{,}000,$
Printing	$3.3P_1 + 3.3P_2 \quad 5.0P_3 + 4.0P_4 \leq 1{,}800{,}000,$
Aging	$2.0P_1 + 2.5P_2 + 2.5P_3 + 1.7P_4 \leq 1{,}500{,}000,$
Dyeing	
(blue)	$2.5D_1 \leq 1{,}500{,}000,$
(red)	$2.9D_2 \leq 1{,}400{,}000,$
Starching	$5.0B + 5.6P_1 + 5.6P_2 + 6.3P_3 + 6.7P_4 + 5.0D_1 + 6.7D_2 \leq 5{,}000{,}000,$
Calendering	$2.5B + 2.0P_1 + 3.3P_2 + 4.0P_3 + 2.5P_4 + 3.1D_1 + 2.9D_2 \leq 4{,}500{,}000.$

Demand Restraints. The demand restraints to be considered are:

$$P_1 \geq 5000,$$
$$P_2 \geq 5000,$$
$$P_3 \geq 5000,$$
$$P_4 \geq 5000,$$
$$P_3 \leq 100{,}000,$$
$$P_4 \leq 50{,}000.$$

Initial Simplex Table

Slack and artificial variables must be introduced to convert the inequalities to equations and provide an initial basic feasible solution. The simplex table for the initial basic feasible solution is indicated in Table 1.

Final Solution

Although this problem is not so large as the preceding coffee problem, it is still advisable to use a computer in solving the problem. The total contribution to profit and overhead for the optimum solution was $1,025,888. The values of the production variables contained in the optimum solution are indicated below:

Solution Variable	*Quantity*
B	—
P_1	5,000
P_2	186,199
P_3	100,000
P_4	50,000
D_1	592,857
D_2	—

It is interesting to note that the optimum production schedule does not include any production of items B and D_2. However, upon showing this schedule to the marketing department, it may well be that the marketing department will insist on the production of some minimum quantities of B and D_2 simply to maintain satisfactory customer relations in markets that require these two products as well as other products sold by the company. In this case a second simplex solution that will include two additional demand restraints, one for B and one for D_2 must be obtained.

The shadow prices of some of the variables not included in the optimum solution (nonbasis variables) are of particular interest. For example, the shadow prices of B and D_2 are \$0.31 and \$0.07 (per unit) respectively. This means that the company must sacrifice \$0.31 of profit for each unit of B it produces and \$0.07 for each unit of D_2 it produces. This information would be quite useful to the marketing department in attempting to decide whether or not any units of B or D_2 should be produced to maintain satisfactory customer relations in certain markets.

The production manager would be interested in the opportunity profit for S_8, which is \$0.35. This represents the marginal value of one unit of mercerizing (dyed) process capacity.[1]

[1] The solution to this problem has been found to be highly sensitive to differences in rounding off input data.

TABLE I

		Column	1	2	3	4	5	6	7	8	9	10	11	12	13	14	15	16	17	18	19	20	21	22	23	24	25	26	27	28	29	30	31	
		C_j	0.40	0.60	0.80	1.00	1.25	1.20	1.30	0	0	0	0	0	0	0	0	0	0	0	0	0	0	0	0	0	0	0	0	$-M$	$-M$	$-M$	$-M$	
Row	C_i	Solution	B	P_1	P_2	P_3	P_4	D_1	D_2	S_1	S_2	S_3	S_4	S_5	S_6	S_7	S_8	S_9	S_{10}	S_{11}	S_{12}	S_{13}	S_{14}	S_{15}	S_{16}	S_{17}	S_{18}	S_{19}	S_{20}	A_1	A_2	A_3	A_4	P_0
1	0	S_1	1.1	1.7	1.1	1.4	1.2	1.1	1.2	1																								1,500,000
2	0	S_2	7.7	10.0	11.1	9.1	12.5	7.7	8.3		1																							15,000,000
3	0	S_3	6.7	11.1	10.0	12.5	11.1	7.7	8.3			1																						9,000,000
4	0	S_4	10.0	9.1	9.5	9.1	9.1	9.1	8.3				1																					15,000,000
5	0	S_5	7.7	10.0	10.0	8.3	9.1	9.1	8.3					1																				14,000,000
6	0	S_6	1.2												1																			830,000
7	0	S_7		1.8	1.7	1.5	1.4									1																		830,000
8	0	S_8						1.4	1.2								1																	830,000
9	0	S_9		3.3	3.3	5.0	4.0											1																1,800,000
10	0	S_{10}		2.0	2.5	2.5	1.7												1															1,500,000
11	0	S_{11}						2.5												1														1,500,000
12	0	S_{12}							2.9												1													1,400,000
13	0	S_{13}	5.0	5.6	5.6	6.3	6.7	5.0	6.7													1												5,000,000
14	0	S_{14}	2.5	2.0	3.3	4.0	2.5	3.1	2.9														1											4,500,000
15	$-M$	A_1		1																				-1						1				5,000
16	$-M$	A_2			1																				-1						1			5,000
17	$-M$	A_3				1																				-1						1		5,000
18	$-M$	A_4					1																				-1						1	5,000
19	0	S_{19}				1																						1						100,000
20	0	S_{20}					1																						1					50,000

The Jones Company

Case

(A Graphical Linear Programming Problem)

Mr. Fred Jones is the owner and manager of a small manufacturing company. His company produces and sells two products—mops and brooms.

The demand for brooms has been very stable, and Jones feels certain that he can sell all of the brooms that he wishes at a net contribution to profit of $1 per unit. The mop market, however, is unstable. Jones estimates that the net unit contribution of mops is likely to be somewhere in the range of $0.50 to $1.50. To plan production, Jones has decided to investigate two cases: Case 1 corresponds to a net unit contribution of mops of $0.50, and Case 2 corresponds to $1.50.

The production processes for mops and brooms can be described as follows: Mops require 1 man-hour of labor and 2 runs on Jones' mop-and-broom machine; brooms require 1 man-hour of labor and 1 run on Jones' mop-and-broom machine. The company has 20 man-hours of labor and a capacity of 20 runs on the mop-and-broom machine.

Mr. Jones must produce at least 10 brooms regardless of profit considerations, because he has a long-standing order with a very important customer for 10 brooms per period. Because of a limitation on raw materials for mops, no more than 5 mops can be produced.

Use the graphical approach to linear-programming to advise Jones on the following points:

1. For Case 1, what is the optimum production schedule and maximum contribution to profit?
2. For Case 1, if Jones could rent 5 more runs on a mop-and-broom machine for \$0.25 per run, should he? What about further increases in runs at the same price of \$0.25 per run?
3. For Case 2, answer the questions in point 1.
4. For Case 2, answer the questions in point 2.

Andrews Investment Service

Case

The Andrews Investment Service has been asked to determine the composition of a $1,000,000 portfolio for a new customer. The customer, Mr. Sasser, is somewhat familiar with the stock market, and as a result, has given Andrews a number of restrictions on how the money is to be invested. For example, Sasser has stipulated that no more than 50 percent can be invested in common stocks, and no less than 40 percent can be invested in bonds. Sasser also requested that Andrews consult him before making the final purchase decision.

After some study, Andrews presented Sasser with a preliminary report. Out of all possible securities on the market, Andrews had selected ten which seemed to be most appropriate for Sasser's investment goals. The ten securities are listed on page 191. The current price and expected return per security are also shown. The expected return, expressed in dollars, includes both expected cash payments (interest and dividends) and expected increases in the price of the security.

Sasser approved the list prepared by Andrews, but he was concerned about the possibility of having too much money invested in airlines. He therefore added a restriction that no more than 15 percent of the portfolio could be in airlines stock. And, as a final stipulation, he requested that at least 25 percent of the portfolio consist of utility company stock (common or preferred).

Preliminary Report for Mr. Sasser

Security Number	*Name*	*Price*	*Expected Return*
1	Power Company (CS)	$ 60	$ 3.50
2	Telephone Company (CS)	50	4.00
3	Life Insurance (CS)	120	8.00
4	Ohio Bonds	1015	45.00
5	New York City Bonds	980	43.00
6	Gas Company (PS)	95	6.00
7	Brown Industries (CS)	24	2.00
8	Dixie Airlines (CS)	70	7.00
9	Northern Airlines (CS)	125	10.00
10	World Airlines (CS)	44	4.00

Note: CS—common stock, PS—preferred stock

Andrews then explained to Sasser how it intended to select the exact composition of the portfolio. Linear programming would be used to choose that portfolio with the maximum total expected return, yet satisfying all of Sasser's restrictions.

Problems

1. Determine the number of shares of each security in the optimum portfolio. What is the total expected return for this portfolio? Interpret the shadow prices of the restraints.

2. Although Andrews' method (with Sasser's help) takes risk into account to some extent, it is primarily concerned with maximizing expected return. For example, of the three airline stocks, the particular stock with the highest return per dollar will be selected. More sophisticated methods have been developed which take the riskiness of each security into account. Thus, a *combination* of the three utility stocks might be preferred to a single utility's stock because of the effect of diversification on total risk. One such method, developed by H. Markowitz, requires quadratic programming.[1] Investigate Markowitz's method and determine what additional data would be required for Mr. Sasser's problem.

[1] H. Markowitz, *Portfolio Selection* (New York: John Wiley, 1959).

The Kraft Pulp & Paper Company

Case

The Kraft Pulp & Paper Company is presently engaged in the production of 5 different grades of paper and paperboard—Chemkraft, Bleached Board, Pine Kraft, Kraft Board, and Newsprint. To produce these 5 different products the company operates 2 large paper machines. These machines may be operated around the clock for 365 days per year.

A sales forecast made by the sales department just prior to January 1, 1971, indicates that expected sales for the company's 5 products for the first quarter of 1971 can be approximated by the schedule shown at top of page 193. Expected demand for each product during each month in the first quarter is expressed in terms of paper machine-hours, the number of paper machine-hours of production capacity required to satisfy the demand for a given product in a particular month. (Although demand for paper products is usually expressed in tons, it is more convenient to utilize conversion factors obtained from the time-study group of the industrial engineering department to express demand in paper machine-hours.)

The expected capacities of the paper mill in machine-hours during the first quarter of 1971 are tabulated below in terms of 3 eight-hour shifts—A, B, and C. Allowances have been made for machine hours consumed by preventive maintenance and breakdowns. Past experience has indicated that breakdowns occur with greater

	Chemkraft	*Bleached Board*	*Pine Kraft*	*Kraft Board*	*Newsprint*
January	100	150	225	—	100
February	250	300	280	200	300
March	300	275	400	300	125

frequency during the two night shifts (B and C). Hence, the paper machine capacities for the B and C shifts are somewhat less than those for the A shift.

	A Shift	B Shift	C Shift
January	450	425	400
February	400	375	350
March	450	425	400

The company's labor contract with the International Brotherhood of Pulp, Sulphite, and Paper Mill Workers stipulates that employees working during the B and C shifts shall receive premium wages. The aggregate labor premium for operating a paper machine during the B shift is $15 per hour, whereas the aggregate labor premium for the C shift is $30 per machine-hour.

The cost accounting department has developed the following schedule of costs for storing 1 machine-hour of production of each of Kraft's 5 products for 1 month.

Product	Storage Cost per Machine Hour per Month
Chemkraft	$8
Bleached Board	$2
Pine Kraft	$4
Kraft Board	$6
Newsprint	$3

Problems

1. Determine an optimum production schedule for the first quarter of 1971.
2. How might one go about estimating the monthly cost of storing one machine-hour of production? What costs would be included in the estimate?

3. In what sense might it be unrealistic to speak of hourly production costs?
4. Discuss several alternative ways of formulating this problem. What effect would these alternatives have on the accuracy of the solution? What effect would they have on the size of the linear programming problem?
5. To what extent is the accuracy of the solution dependent on the accuracy of the sales forecast?

South American Airways

Case 6

South American Airways is a nonscheduled cargo airline that operates in 5 cities in the United States and throughout Central and South America. The airline operates a fleet of 12 aircraft consisting of four C-46's, two DC-4's, and six DC-6's. These aircraft are used to pick up and deliver cargo whenever orders are received from customers.

One of the major problems facing the traffic manager of South American Airways is that of scheduling the pick up and delivery of cargo. On a particular day the traffic manager receives the following information showing the present location, final destination and weight (in pounds) of 12 customer orders that are to be shipped by the airline within the next 24 hours.

Present Location	*Final Destination*	*Weight (in pounds)*
Atlanta	San Juan	25,000
Caracas	Santiago	10,000
Havana	Miami	10,000
Houston	San Juan	15,000
Mobile	Rio de Janeiro	10,000
Monterrey	Mexico City	15,000
New Orleans	Quito	20,000
Rio de Janeiro	New Orleans	15,000
San Salvador	New Orleans	10,000
Sao Paolo	Buenos Aires	10,000
Santiago	Caracas	10,000
Tampa	Guatemala City	20,000

The location of each of the airline's 12 aircraft, by type, is shown below.

City	*Type of Aircraft*
Bogota	C-46
Buenos Aires	DC-6
Caracas	C-46
Guatemala City	DC-4
Mexico City	DC-4
Miami	DC-6
New Orleans	C-46
New Orleans	DC-6
Rio de Janeiro	C-46
Rio de Janeiro	DC-6
San Juan	DC-6
San Salvador	DC-6

The table below indicates for each of the cargo loads available the maximum ranges (in miles) that can be flown before refueling is required on each of the 3 different types of aircraft. The data indicates that the maximum loads for the C-46, DC-4, and DC-6 are 15,000 pounds, 20,000 pounds, and 25,000 pounds respectively

Cargo Weight	*C*-46	*DC*-4	*DC*-6
Empty	1200	2800	3200
10,000 pounds	900	2800	3200
15,000 pounds	200	2800	3200
20,000 pounds	—	1200	3200
25,000 pounds	—	—	1000

The optimum cruising speeds for the airline's aircraft are as follows.

Aircraft	*Optimum Speed*
C-46	170 m.p.h.
DC-4	200 m.p.h.
DC-6	300 m.p.h.

Table 1 indicates (1) the approximate distances (in miles) between the present location of aircraft and the present location of cargo orders, and (2) the approximate distances between the present locations of cargo orders and the final destination of cargo.

TABLE I

	Bogota	*Caracas*	*Buenos Aires*	*Guatemala*	*Mexico City*	*Miami*	*New Orleans*	*Rio de Janeiro*	*San Juan*	*San Salvador*	*Santiago*
Atlanta	1700	1600	4000	1150	1100	500	400	3800	1300	1200	3750
Caracas	500	—	2450	1250	1750	1100	1650	2150	450	1200	2400
Havana	1100	1050	3350	650	900	200	600	3200	900	650	3100
Houston	1800	1850	4000	900	600	850	300	3950	1700	550	3700
Mobile	1650	1600	3900	900	850	900	125	3750	1400	1000	3650
Monterrey	1800	1950	3950	800	350	1050	600	4000	1850	1000	3600
Quito	350	850	2200	1000	1100	1450	1750	2250	1200	950	1900
Rio de Janeiro	2200	2150	1100	3200	3700	3700	3800	—	2500	3100	1600
San Salvador	1150	1200	3100	80	600	850	950	3100	1250	—	2750
Sao Paulo	2100	2100	900	3100	3550	3200	3700	200	2450	3000	1400
Santiago	2100	2400	650	2850	3250	3300	3600	1600	2800	2750	—
Tampa	1350	1300	3650	850	1000	200	400	3450	1050	900	3850

The 21 airports mentioned in the table above can accommodate all types of piston-engine aircraft including the 3 types of aircraft that South American Airways presently uses. In addition to these airports there are 8 other airports in the Latin American area the airline serves that have runways of sufficient lengths for the company's aircraft to land and take off safely. These are: San Jose, Managua, Panama, Teguci, Lima, Kingston, Ciudad Trujillo, and Belem. Any of these 27 airports may be used for intermediate stops or flights requiring refueling operations. Although DC-4's and DC-6's can land only at the aforementioned airports, C-46's can land almost anywhere, provided there is at least 2000 feet of runway. Hence, the number of possibilities for C-46 intermediate landings is extremely large. (Intermediate airports may be determined from a world atlas or from material published by the Civil Aeronautics Board.)

The airline has computed the average operating costs per hour for each of its 3 types of aircraft. This data is tabulated below for fully loaded aircraft.

Costs	*C-46*	*DC-4*	*DC-6*
Crew Salaries	$23	$23	$30
Crew Expenses	3	3	3
Fuel	40	50	80
Oil	2	2	3
Hull Insurance	2	4	8
Liability Insurance	1	1	1
Maintenance			
Labor	20	30	45
Materials	5	10	15
Engine Overhaul	20	25	40
Air Frame Overhaul	3	5	10

The operating costs for an empty aircraft are identical to those of a loaded aircraft with the exception of the costs of fuel and oil.

Costs	*C-46*	*DC-4*	*DC-6*
Fuel	$35.00	$45.00	$75.00
Oil	$ 2.00	$ 1.75	$ 2.50

Landing charges vary rather widely among airports and are usually a function of the weight of the particular aircraft in conjunction with a multiplicity of other variables. The approximate landing charges[1] for the major airports in Latin America are shown in the table at top of page 199 for each of the airline's 3 types of aircraft. The average landing fee for C-46 aircraft in Latin America (including airports whose runways are less than 3000 feet in length) is around $20.

[1] Although these landing charges are fictitious, they do reflect the approximate range of actual landing charges.

Airports	*C*-46	*DC*-4	*DC*-6
Atlanta	$25	$40	$75
Belem	15	30	50
Bogota	20	30	35
Caracas	30	35	50
Ciudad Trujillo	40	55	70
Buenos Aires	30	40	60
Guatemala City	45	65	90
Havana	45	65	90
Houston	20	20	30
Kingston	25	35	50
Lima	25	30	40
Managua	30	50	60
Mexico City	15	20	25
Miami	20	30	40
Mobile	20	25	35
Monterrey	10	15	20
New Orleans	30	45	80
Panama	15	20	30
Quito	10	10	15
Rio de Janeiro	30	35	50
San Juan	15	25	25
San Jose	20	25	25
San Salvador	20	20	30
Santiago	20	30	40
Sao Paulo	25	25	35
Tampa	30	40	65
Teguci	25	30	45

Problems

1. With the aid of an atlas, formulate a linear programming model of the problem facing the traffic manager assigning aircraft to cargo flights. A cargo flight is defined as the movement of an aircraft from its present location to a pickup location and finally the delivery of the cargo to the final destination.
2. Formulate this problem by the special linear programming model known as "the assignment problem."
3. Solve the problem by the transportation method and by the assignment method.
4. What are the relative advantages and disadvantages of each method of solution?

The Frosty Frozen Food Company

The Frosty Frozen Food Company, a distributor of frozen vegetables, serves over 300 retail grocery stores in a large metropolitan area. Frosty distributes 15 different vegetables that are packed in cartons with 12 ($5\frac{1}{4}$ in. $\times$ 4 in. $\times$ $1\frac{5}{8}$ in.) packages to a carton.

The company replenishes its supply of frozen foods only once each week. With the exception of the summer months, when fresh vegetables are available to consumers, the company rarely has any inventory remaining in the warehouse at the end of a week. An entire week's supply of frozen vegetables arrives each Monday morning at the company's cold-storage warehouse. Although the warehouse contains 16,000 cubic feet of space (40 ft $\times$ 40 ft $\times$ 10 ft), frozen vegetables may be stacked only to a maximum height of 6 feet. This restriction is imposed by overhead refrigeration equipment and the possibility of damaging the frozen food packages by the excessive weight placed on those cartons that are on the bottom of the stack. The warehouse also contains 2 five-foot aisles that are parallel to the length of the building and 2 five-foot aisles that are parallel to the width of the building. The aisles must remain clear at all times for efficient materials handling.

Frosty's supplier of frozen foods extends to the company a line of credit amounting to $40,000 per week. That is, the company's weekly credit purchases from its supplier may not exceed $40,000. Additional purchases must be financed through short-term bank loans, since the company feels that it must maintain its present cash balances in order to preserve its liquidity position. The extent to which the company relies on bank loans depends on the prevailing interest rates and the demand conditions for frozen vegetables.

The following table represents the predicted sales (in 12-package cartons) for the company's 15 products during a forthcoming week. This forecast is expressed in terms of the maximum and minimum expected sales for the particular week. The minimum sales figure is based on a contractual agreement that Frosty has with a large retail grocery chain to provide certain minimum quantities of certain designated products each week. The maximum figure represents an estimate of the expected number of cartons of each vegetable that can be sold during the week in question.

	Minimum	*Maximum*
Whipped potatoes	300	1500
Creamed corn	400	2000
Black-eyed peas	250	900
Artichokes	—	150
Carrots and peas	300	1200
Succotash	200	800
Okra	150	600
Cauliflower	100	300
Green peas	750	3500
Spinach	400	2000
Lima beans	500	3300
Brussels sprouts	100	500
Green beans	500	3200
Squash	100	500
Broccoli	400	2500

The cost and selling price per carton of each of the products in the company's product line are listed at top of page 202.

	Cost/unit	*Selling Price/unit*
Whipped potatoes	$2.15	$2.27
Creamed corn	2.20	2.48
Black-eye peas	2.40	2.70
Artichokes	4.80	5.62
Carrots and peas	2.60	2.92
Succotash	2.30	2.48
Okra	2.35	2.20
Cauliflower	2.85	3.13
Green peas	2.25	2.48
Spinach	2.10	2.27
Lima beans	2.80	3.13
Brussels sprouts	3.00	3.18
Green beans	2.60	2.92
Squash	2.50	2.70
Broccoli	2.90	3.13

Problems

1. Assuming the company does not utilize its present cash balances or a short-term bank loan to finance its frozen food purchases, determine by linear programming the optimum quantity of each product that the company should order for the forthcoming week.
2. What is the maximum interest rate that Frosty should consider paying in order to increase its purchases beyond the $40,000 limitation imposed by the supplier? If the company can obtain a short-term note for less than this maximum interest rate, what effect will this have on the solution?
3. Suppose the company can lease additional cold storage space near its present location. Using the assumptions in Question 1 and then Question 2, determine the maximum rental fee (per cubic foot of cold storage capacity) that the company can afford to pay each week for increased storage capacity.

The Natchez Oil Company[1] *Case 8*

The Natchez Oil Company produces and distributes two different types of gasoline, "regular" and "premium." These gasolines must meet the following specifications:

Regular Gasoline Specifications

Vapor pressure index	$\leq$ 26
Performance number	$\geq$ 61.4
Performance number	$\leq$ 62.0
Fraction distilling off at 150°F	$\leq$ 32%
Fraction distilling off at 200°F	$\geq$ 45%
Fraction distilling off at 250°F	$\geq$ 64%
Fraction distilling off at 350°F	$\geq$ 90%

Premium Gasoline Specifications

Vapor pressure index	$\leq$ 26
Fraction distilling off at 150°F	$\leq$ 40%
Fraction distilling off at 200°F	$\geq$ 56%
Fraction distilling off at 300°F	$\geq$ 90%

[1] The data for this problem was obtained by permission from Kurt Eisemann's *Linear Programming* (New York: I.B.M. Applied Science Division, January 1954), pp. 37–38. © 1954 by International Business Machines Corporation.

On a particular day the company has 8 different gasoline components (C_1, C_2, C_3, C_4, C_5, C_6, C_7, C_8) available for blending into regular and premium gasolines. These components have the following technical characteristics:

Characteristics	C_1	C_2	C_3	C_4	C_5	C_6	C_7	C_8
Contribution to vapor pressure index	12	24	0	28	8	0	16	140
Performance number (analogous to octane number)	56	58	57	66	68	66	74	74
Fraction distilling off at 150°F (percent)	8	20	0	91	11	0	7	115
Fraction distilling off at 200°F (percent)	39	54	0	110	50	3	26	100
Fraction distilling off at 250°F (percent)	77	88	5	100	74	6	62	100
Fraction distilling off at 300°F (percent)	96	100	25	100	88	20	82	100
Fraction distilling off at 350°F (percent)	100	100	90	100	100	52	95	100

(*Note:* Percentages greater than 100% are due to the "carry" effect—distilling vapors carry along other, less volatile components.)

The maximum availabilities of the 8 gasoline components in barrels per day are tabulated below:

Component	*Availability*
C_1	11,500 barrels
C_2	5,000
C_3	1,500
C_4	5,500
C_5	6,500
C_6	4,750
C_7	1,000
C_8	unlimited

The contributions to profit and overhead (per barrel of gasoline component utilized) are shown below:

Component	*Contribution*
C_1	\$0.448
C_2	0.464
C_3	0.456
C_4	0.528
C_5	0.544
C_6	0.528
C_7	0.592
C_8	0.592

Problems

1. The company is presently realizing profits of \$15,000 per day by blending its gasolines by trial-and-error and intuitive means. Should the company consider using linear programming to blend its gasolines if the costs of processing the required data and solving the problem on a computer amounts to \$250 per day?
2. Use linear programming to determine the optimum quantities of each gasoline component to blend into regular and premium gasolines.

The Midwest Grain Elevator

Cas

The Midwest Grain Elevator is owned and operated by a large midwestern city. The major functions of this publicly owned elevator are to receive, store, and ship grain. The elevator charges a service fee for both handling and storing grain. In general, it can be said that the major objective of the Midwest Grain Elevator is to receive, store, and ship as much grain as possible.

Grain is delivered to the elevator by the owners of the grain, who are usually referred to as "sellers." The grain is then stored in the elevator until a "buyer" can be found, or until the "buyer" is ready to receive the grain. When one of the elevator's customers (a "seller") receives an order for a certain type of grain, it is the responsibility of the elevator to prepare for shipment that quantity of the "seller's" grain (stored in the elevator) which will satisfy the requirements of the order placed by the "buyer." The order may be satisfied by selecting only one type of grain from the "seller's" grain supply, or by blending several different grades in order to obtain a grain that will satisfy the requirements of the order.

The elevator does not have a direct interest in the cost of grain, because its revenues are derived from service charges and not from buying and selling; however, it does have an indirect interest in the cost of a particular blend. If the "buyers" of the grain are satisfied with the cost of the grain they purchase, then they may continue to purchase grain through the Midwest Elevator. However, it they feel that other elevators can provide the same grain at a lower cost, then there may be a tendency for "buyers" to go to other elevators to make their purchases. Hence, there is a definite incentive for the Midwest Grain Elevator to provide optimum (minimum cost) blends to satisfy the requirements of the orders received by its customers. If the grain elevator provides the optimum blend to fulfill customer orders, the owner of the grain ("seller") also benefits. He will be able to satisfy more high-quality orders with the grain he stores in the elevator. Thus, the grain elevator is able to increase the quality of the service it offers to both "buyers" and "sellers."

On a certain day one of the elevator's customers, The Saw Mill River Feed & Grain Company, had 497,900 bushels of corn stored in the elevator. Obviously, it would be impossible to identify the exact kernels of corn that The Saw Mill River Feed & Grain Company delivered to the elevator. Hence, Table 1 simply represents

TABLE I

Type of Corn	*Quantity (bushels)*	*Cost per Bushel*	*% Moisture Content*	*Weight per Bushel (lbs.)*	*% Total Damage*	*% Foreign Material*
1	15,000	$1.50	11	58	3	1
2	30,000	1.45	12	57	2	1.5
3	45,000	1.44	15	57	2	1
4	25,000	1.45	12	58	3	3
5	40,000	1.42	13	56	4	2
6	20,000	1.38	15	54	4	2
7	30,000	1.37	15	55	5	3
8	75,000	1.37	18	57	5	1
9	15,000	1.39	14	58	2	4
10	50,000	1.39	15	55	3	2
11	16,000	1.27	17	53	7	5
12	20,000	1.28	15	55	8	3
13	10,000	1.17	22	56	9	5
14	12,000	1.22	18	54	13	5
15	6,000	1.12	17	55	20	8
16	2,000	1.18	13.5	57	30	2
17	5,000	1.42	13.5	57	3	2
18	4,000	1.42	13.5	57.5	3	2
19	5,000	1.42	13.5	57	5	3
20	6,000	1.36	15	56.5	10	4
21	1,300	1.29	13.5	58	15	4
22	29,000	1.29	13	57	15	5
23	1,900	1.42	13	58	3	2
24	33,000	1.40	13	56	5	3
25	6,700	1.22	14	56	20	5
26	5,000	1.30	15	56.5	10	4

the amounts of different types of corn credited to Saw Mill River's account with the elevator. The 497,900 bushels are segregated into 26 different types of corn, which differ according to 1. quantity available, 2. cost per bushel, 3. percentage moisture content, 4. weight per bushel, 5. percentage damaged, and 6. percentage foreign material.

The Saw Mill River Feed & Grain Company has received orders for 6 different grades of corn, which are to be supplied by the elevator from the corns that Saw Mill River has stored there. The quantity of each of these 6 grades, as well as the maximum moisture content, minimum weight per bushel, maximum percentage damaged, and maximum percentage foreign material, are given in the following table.

Grade of Corn	*Quantity (bushels)*	*Maximum % Moisture*	*Minimum Wt./bu. (lbs.)*	*Maximum % Damage*	*% Foreign Material*
1	40,000	13	56	2	2
2	32,000	15.5	54	5	3
3	50,000	15	56	2	4
4	20,000	16	54	9	4
5	30,000	23	54	10	6
6	100,000	20	54	9	4

Problems

1. Use linear programming to determine optimum blends of available corns to use in satisfying the 6 orders received by The Saw Mill River Feed & Grain Company.

Ling Chemical Company *Case 10*

Ling Chemical Company is a major producer of bulk chemicals such as sodium chloride, calcium chloride, chlorine, and phenol, as well as various agricultural chemicals used in the production of fertilizers and insecticides. These chemicals can be produced at each of Ling's 3 production facilities, which are located in Bay City, Michigan, Baton Rouge, Louisiana, and Galveston, Texas. Although these products may be shipped directly to the customer from any of the 3 production facilities the company finds that it is frequently more economical to ship via one of its 4 shipping terminals and thus take advantage of bulk shipping rates that are made available by railroads, barge lines, and steamship lines. Ling's storage terminals are located in Chicago, St. Louis, Charleston, and New York.

For scheduling production at the 3 plant locations, Ling utilizes a monthly planning period. In other words, the company makes monthly forecasts of the demand for each of its products for the forthcoming month. Although the company ships to over 2000 customer locations, these have been consolidated into 316 customer locations to reduce the scope of the problem. (It is assumed that a minimum of accuracy is lost in condensing the data according to this technique, for customers are simply grouped into locations having comparable freight rates.) In addition to forecasting monthly demand by product for each customer loation, Ling also estimates the available capacity at each plant for each product which is to be produced.

Cost accounting data is available at each of the 3 plants for estimating unit production costs for each product. The traffic department can provide shipping rates per unit for each product. This information is available for all 15 possible shipping patterns.

Plant	*Terminal*
Bay City	Direct
Bay City	Chicago
Bay City	St. Louis
Bay City	Charleston
Bay City	New York
Baton Rouge	Direct
Baton Rouge	Chicago
Baton Rouge	St. Louis
Baton Rouge	Charleston
Baton Rouge	New York
Galveston	Direct
Galveston	Chicago
Galveston	St. Louis
Galveston	Charleston
Galveston	New York

Shipping rates from a production plant to a customer location via a storage terminal include both the rate from the plant to the storage terminal and the rate from the terminal to the customer.

Problems

1. Using linear programming, develop a technique for determining the optimum allocation of production to the available production and shipping facilities.
2. Discuss some of the difficulties that may arise in accumulating the data for this problem.
3. Assume that the St. Louis, Chicago, Charleston, and New York shipping terminals have only limited storage capacities. What effect will this have on your solution?
4. This problem may also be framed in terms of the special case of linear programming known as the "transportation method." What are the computational advantages and disadvantages of framing the problem according to the transportation technique as opposed to solving it by the simplex method?
5. Which computers would be best suited to solving a problem of this magnitude? (Consult the current literature.)

The Wilcox Manufacturing Company

Case 11

The Wilcox Manufacturing Company produces an item that is subject to wide seasonal fluctuations in sales. The fluctuations in the demand for this product are illustrated by the sales forecast chart (Figure 1) for the forthcoming year.

The firm must always meet the monthly demand requirements or risk the possibility of losing customers. It can fulfill individual demands either by producing the required amount during the particular month or by producing only part of the required amount and making up the difference by using the overproduction (inventory) from previous months.

There are many different production schedules that the company could consider. For example, it could produce the exact number of units required by the sales forecast for each month. But this would involve a widely fluctuating production pattern that would be costly to maintain because of overtime costs and hiring costs in high-production months, and because of layoff costs (personnel and machinery) during low-production months. Another alternative available to Wilcox would be to overproduce in periods of low requirements, store the surplus, and use the excess in periods of high requirements. This would result in a rather stable production pattern, but perhaps a costly one because of the cost of storing the surplus. Such a solution

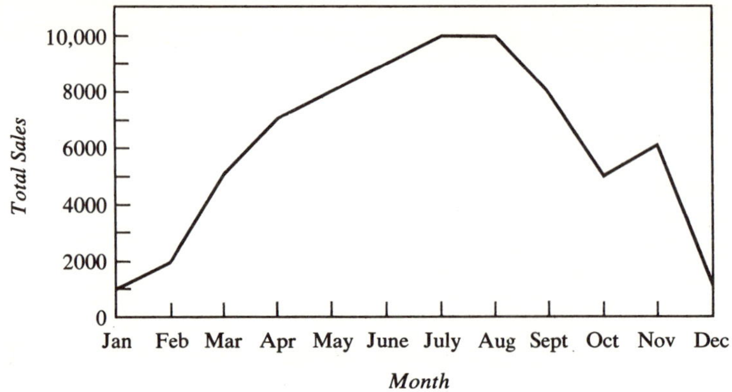

Figure 1. Sales Forecast Chart.

would be most undesirable if it yields relatively large monthly surpluses. Neither of these extreme cases would be satisfactory to management since its objective is to determine a production schedule that minimizes the sum of the costs due to output fluctuations and to inventories.

The cost accounting department has estimated that it costs approximately \$1.00 to increase output by 1 unit from one month to another and \$0.50 to reduce output by 1 unit from one month to another. The monthly cost of storing 1 unit of product is \$0.40. The production schedule for the month of December in the current year calls for the production of 1000 units, and it is estimated that the inventory level on January 1 will be 1500 units.

Problems

1. Determine an optimum production schedule for the Wilcox Company for the forthcoming year.
2. What costs might be included in the marginal cost of increasing output and the marginal cost of decreasing output for a manufacturing firm? Why might these marginal costs be difficult to determine for many firms?
3. Suppose the marginal costs of increasing and decreasing production cannot be determined exactly; what alternatives are available to the production manager, if he wishes to use linear programming to find an optimum production schedule?

The Wabash Refining Company *Case 12*

The Wabash Refining Company operates two small petroleum refineries, one located in St. Louis, Missouri, and the other in Evansville, Indiana. The St. Louis refinery receives its crude oil via a pipeline that originates in the oil fields of East Texas. The Evansville refinery purchases crude oil from bulk terminals located in the Southern Illinois oil-producing region.

"Regular" gasoline accounts for over 50% of the company's output from a barrel of crude oil input. The St. Louis refinery has a rated output of 20,000 barrels of "regular" gasoline per day, whereas the Evansville refinery has a rated output of only 5000 barrels per day.

DISTRIBUTION SYSTEM

The principal channel of distribution utilized by Wabash consists of 15 bulk storage terminals located in the company's three-state marketing area, Illinois, Indiana, and Missouri. These bulk terminals receive their supply of gasoline by tank car, barge, and pipeline. The company operates two gasoline pipelines. One of these lines connects the St. Louis refinery with Chicago and the other runs between Evansville and Indianapolis. The 15 bulk terminals that the company presently operates are listed on page 172. Some of these terminals are owned by the company, but others are leased. (Bulk storage facilities are also available at the St. Louis and Evansville refineries.)

The gasoline is distributed from the bulk terminals by company-owned trucks to service stations, farms, industrial plants, commercial accounts, city, state, and federal agencies. The area served by a bulk terminal varies considerably, depending on such things as population density and road conditions. Although the radius of the area served by a terminal is usually not in excess of 50 miles, three of the terminals serve areas with a radius greater than 150 miles.

Illinois	*Indiana*	*Missouri*
Chicago	Fort Wayne	Cape Girardeau
Danville	Indianapolis	Jefferson City
Peoria	Muncie	Kansas City
Springfield	South Bend	Portageville
	Terre Haute	St. Joseph
		Springfield

The Wabash Refining Company owns 325 service stations that it leases to dealers who operate these stations. In addition to the 325 service stations that it owns, Wabash has over 800 other accounts in the three-state area.

DISTRIBUTION COSTS

In recent years the company has enjoyed a rather substantial increase in the demand for its "regular" gasoline, but its profits on gasoline operations have declined. Management feels that profits have failed to increase because of the excessive transportation costs associated with an insufficient number of bulk terminals in the three-state area. The long truck hauls required to transport gasoline from some of the bulk terminals to customers in the remote sectors of the area served by the terminal have proved to be extremely costly.

Management is considering the possibility of increasing the number of bulk terminals used for distributing gasoline, but they are not certain just how many terminals to operate, where to locate them, or how to build them. Management's problem is further complicated by the fact that the demand for gasoline is subject to rather wide seasonal variations. This is indicated in the table on page 215, which shows average daily gasoline production in barrels per day for each month in the preceding year's operations. (This includes production at both refineries.)

Month	*Barrels per Day*
January	15,000
February	15,000
March	17,000
April	20,000
May	20,500
June	23,000
July	24,000
August	23,500
September	22,000
October	19,000
November	18,000
December	18,000

OBJECTIVES OF MANAGEMENT

Specifically, management would like to have the answers to the following questions:

1. How many bulk terminals should it operate in order to minimize total distribution costs?
2. Where should these terminals be located?
3. How large should the terminals be?
4. Which one of the two refineries should serve the different terminals in order to minimize total production and distribution costs?

Although actual cost and demand data are not required to set up the following problems, if the reader desires to obtain practical experience with the manipulation of a problem of this magnitude it may be desirable to assume a set of production and distribution costs, as well as individual customer demands, and solve the problem on an electronic computer.

Problems

1. With the aid of last year's monthly sales data, develop a linear programming mode to simulate the company's gasoline market and determine the optimum number, size, and location of the bulk terminals that should have been operated in each month of the preceding year.
2. Using the information in Question 1 as well as your own judgment, determine the overall optimum number, size, and location of bulk terminals that should have been operated last year.

The Ward Advertising Case

Case 13

Professor Scott Ward, a professor of marketing at a leading business school, decided to put into practice some of the scientific methods that he was teaching his students. He firmly believed that advertising was an area where his know-how in operations research methods would pay off. Ward's impression was that most advertising decisions were made on the basis of rules-of-thumb, or rough judgments. Hence, he believed that he would be deluged by prospective clients if he could demonstrate the value of the modern management science approach to a few large advertisers.

Professor Ward's first opportunity came as a result of a lecture he gave to a group of beer marketing executives. After the talk, John Parrish of Schultz Brewery told Ward of some difficulties that Schultz was having in planning their advertising campaigns. Specifically, Mr. Parrish and some of the other members of the Schultz management felt that the company was not getting the maximum effectiveness from their advertising expenditures. However, Mr. Parrish continued, they did not know how they might improve the situation. Ward then outlined his ideas to Mr. Parrish and was immediately retained to prepare a proposal for a Schultz advertising campaign. Ward's role would be to recommend an allocation of advertising expenditures among various media, yet he would not be involved in the actual content of the advertising.

Schultz, a national brewer, distributed beer in 42 large cities. Since Schultz beer was regarded as a premium beer, it appealed to only a particular class of beer drinkers. Thus, it was believed to be important to consider not only the *total* audience reached by a medium, but also the *composition* of the audience (that is, the percentage of males and females, income classes, age classes, and so forth).

Professor Ward also stressed two other factors that should be taken into account in evaluating media: (1) the percentage of the relevant audience that would probably be exposed to an ad—for example, not all of the potential Schultz beer consumers who read a magazine also read a particular ad in that magazine—and (2) the estimated quality of the ad in terms of its expected effectiveness—for example, a page in magazine *A* might be expected to have a greater effect on sales than a page in magazine *B*.

Thus, Ward decided to construct an index of "effective exposure," or EE, for each alternative type of advertising. Of course, in constructing the index he relied heavily on the experience and judgment of the members of Mr. Parrish's staff.

To explain the EE index to Mr. Parrish, Ward selected a full page advertisement in magazine *A* as an example. The total audience was estimated to be 6,000,000. However, after considering the composition of the total audience, it was estimated that 4,000,000 was a better figure to represent potential Schultz customers. Next, Ward applied two other factors to further reduce the number of potential customers. It was estimated that only about 35 percent of the 4,000,000 would actually read the ad in magazine *A*, or $(0.35)(4{,}000{,}000) = 1{,}400{,}000$. Second, the qualitative index of magazine *A* was adjudged to be 0.7 relative to other media (on a scale ranging from 0 to 1.0). Thus, the EE index for a page in magazine *A* was computed as $(0.7)(1{,}400{,}000)$, or 980,000.

The complete list of types of advertising that Schultz could use is given below. The cost per unit and Ward's EE index are also included in the list.

Mr. Parrish was impressed with Professor Ward's method for quantifying the exposure effectiveness of the various media. He remarked to Ward, however, that there were a number of other things that had to be considered, and it would not be possible simply to choose those alternatives that yielded the highest effective exposure per dollar of cost. For example, the Schultz management believed it essential to place at least one full-page ad in at least one issue of both magazines *A* and *C* over a six-month period. Furthermore, management had firm beliefs about the need to allocate at least half of the advertising budget to magazine advertising, and to sponsor at least one baseball game on television.

Ward assured Mr. Parrish that such considerations could be handled easily within a linear programming framework. He then gave Mr. Parrish a brief explanation of the principle of linear programming, explaining that the objective was to maximize the total effective exposure subject to various restraints. He pointed out that the restraints could be judgmental, budgetary, or otherwise. For example, since his proposal was to be for a six-month period, restraints would be needed to prevent

ALTERNATIVE TYPES OF ADVERTISING AVAILABLE TO SCHULTZ

Type		Cost per Unit	Effective Exposure per Unit (EE)
Magazines			
(Weekly) A	page	$60,000	980,000
	half-page	35,000	512,000
(Weekly) *B*	page	20,000	230,000
(Monthly) *C*	page	12,000	140,000
(Weekly) *D*	page	13,000	215,000
	half-page	8,500	112,000
(Monthly) *E*	page	10,000	110,000
	half-page	6,000	96,000
(Weekly) *F*	page	5,500	78,000
Newspapers			
1000-line advertisements in 42 cities, per week		33,000	570,000
Television			
Network-baseball game sponsorship		125,000	1,185,000
Spot-two commercials in 42 cities, per week		18,000	238,000

physical impossibilities—such as a solution calling for 30 full-page ads in magazine *A* when there could only be 26 issues in the period (Schultz never advertised on more than one page in any single issue).

Problems

1. Given a six-month advertising budget of $525,000, what is the best advertising program for Schultz?
2. Suggest some sensitivity analysis studies that should be made.

Maluland

(An Input-Output Problem with Choice)

The King of Maluland is concerned about using the country's resources efficiently. There are two industries in Maluland: industry C produces consumption goods and industry K produces capital goods. The final demand goods, or final bill of goods, for C and K are denoted by D_c and D_k, respectively. The King and his economic advisors have agreed that the relative values of D_c and D_k should be:

D_c at 0.65 malu-dollars/unit,

D_k at 1.00 malu-dollars/unit.

A recent survey that the King ordered shows the total resources of Maluland to be:

800,000 man-hours of labor,

95,000 acres of land,

200,000 units of capital.

The latest input-output table of Maluland implies the following table of technical coefficients (that is, the A matrix of Chapter 9):

Technical Coefficients in Maluland

	C	K
C	0	0.5
K	0.25	0

Engineers in the consumption-goods industry and the capital-goods industry have provided the following data: to produce one unit of C requires 3.5 man-hours, 0.5 acres, and 0.625 units of capital; one unit of K requires 2.5 man-hours and 1.25 units of capital.

The King is interested, therefore, in maximizing the value of net output in Maluland,

$$Z = 0.65D_c + 1.0D_k,$$

subject to the production restraints and the resource availability restraints. Set up the linear programming problem and answer the questions below:

Problems

1. What is the optimum final bill-of-goods in Maluland?
2. What are the total outputs of C and K implied by the answer to Problem 1?
3. What are the shadow prices, or imputed values, of the resources in Maluland? If the shadow price of one of the resources is zero, what is the value of its associated slack variable?

Hint: The production restraints are given by equation (5) of Chapter 9.

The Key Largo Shipyard

Case

(A Two-Person Zero-Sum Game)

The management of the Key Largo Shipyard will soon be negotiating a new contract with the union that represents the company's employees, the International Brotherhood of Ship Carpenters, Caulkers, and Joiners. (The union's contract is renegotiated each year.) The one issue involved at this time is the wage rate. As has been the case for the past twenty years, both the company and the union have agreed to submit their dispute to an arbitration board.

The tripartite arbitration board consists of 3 members, a company representative, a union representative, and an impartial member appointed by the American Arbitration Association who acts as chairman.[1] In most cases the final decision is usually made by the chairman, since the union representative seldom feels compelled to vote with management and vice-versa.

The selection of a management representative is of utmost importance since his actions may well influence the final outcome of the dispute, even though the chairman is actually impartial. The chairman's decision may well be influenced by the

[1] The American Arbitration Association maintains a national panel of more than 13,000 persons, each of whom is an expert in his chosen field or profession and qualified to render impartial decisions in arbitration cases.

choices made by both the union and management as to their respective representatives. For example, if management selects as its representative an individual who is a radical company man, and the union chooses an individual who is either unbiased or does not express his favoritism toward the union vocally, the chairman may have a negative reaction against the company and favor the union in his final decision.

Management is considering 5 different types of individuals to represent the company on the arbitration board. These individuals are indicated below by the symbols M_1, M_2, M_3, M_4, and M_5.

Management Representative	*Characteristics*
M_1	An unbiased individual.
M_2	An individual whose voting record in arbitration cases indicates a slight bias in favor of management, but who on occasions has voted against management.
M_3	A company man who utilizes a "soft-sell" approach.
M_4	An outspoken company man.
M_5	A radical company man.

The union has a similar set of individuals whom it is considering to represent it in the dispute. The individuals are indicated below by U_1, U_2, U_3, U_4, and U_5.

Union Representative	*Characteristics*
U_1	An unbiased individual.
U_2	An individual whose voting record in arbitration cases indicates a slight bias in favor of the union, but who on occasions has voted against the union.
U_3	A union man who utilizes a "soft-sell" approach.
U_4	An outspoken union man.
U_5	A radical union man.

Past experience in arbitration cases has indicated that the average hourly wage increase (in cents per hour) associated with the type of representative that management employs at the bargaining table is completely dependent on the type of representative that the union chooses. And likewise, the average hourly wage increase (in cents per hour) associated with the type of representative that the union employs is dependent on the type of representative that management chooses. (It is convenient to refer to the choices available to the union and management as strategies.)

It should be emphasized that neither the union nor management knows the other's final strategy decision until the arbitration hearing actually begins. Neither party can alter its choice of a representative once the representative's name has been submitted to the chairman of the arbitration board.

Twenty years of arbitration experience indicates that the following table is an *approximate* indication of the expected average wage increases (in cents per hour) associated with the different possible combinations of union and management representatives chosen to serve on the arbitration board.

TABLE I

	M_1	M_2	M_3	M_4	M_5
U_1	0	5	10	10	30
U_2	10	5	15	15	35
U_3	25	15	10	20	40
U_4	15	5	5	15	20
U_5	−5	0	5	10	10

Although the average wage increases in this table are not an exact representation of the wage increases associated with the different union-management strategies utilized over the past twenty years, management does feel that it will prove to be a useful guide in selecting the type of individual to represent the company in future wage negotiations. Furthermore, both the union and management know the approximate wage increases associated with each union-management strategy combination.

Although management would like to select that strategy which will minimize the average hourly wage increase during the forthcoming contract period, it is more concerned with selecting a set of strategies that will enable it to minimize the increase in wage rates over an extended period of time. In other words management would like to select the particular (pure) strategy that will minimize wage increases in the long run. Management may find that its optimum strategy is one in which it does not choose the same particular (pure) strategy each year. Instead, it may select strategies randomly according to some predetermined optimal probability distribution imposed over its pure strategies (M_1, M_2, M_3, M_4, and M_5). Such a strategy is called a mixed strategy.

The union is searching for an optimal pure strategy or mixed strategy that will maximize wage increases over an extended period of time.

Problems

1. What assumptions are necessary to express this bargaining problem as a two-person zero-sum game? Are these assumptions realistic?
2. According to Table 1, which three strategies will not be used at all by management? Why?
3. Which three strategies will not be chosen by the union? Why?
4. Assuming that Table 1 is an approximate representation of the union-management bargaining situation for the next 10 years, use linear programming to obtain an approximation to the optimum mixed strategy for management. What will be the expected wage increase during each contract period using this optimum mixed strategy for the next 10 years?
5. What is the union's optimum mixed strategy?
6. If the assumptions in 1 and 4 are not completely realistic, is this approach to union-management bargaining problems invalidated? Why?
7. How would management decide which five individuals to consider?
8. How would management know what the union was considering?
9. How would past experience in union-management negotiations be organized to generate a payoff table such as Table 1?
10. Is it really appropriate for management to focus so heavily on minimizing long-run wage increases?

Index